SpringerBriefs in Physics

SpringerBriefs in Physics are a series of slim high-quality publications encompassing the entire spectrum of physics. Manuscripts for SpringerBriefs in Physics will be evaluated by Springer and by members of the Editorial Board. Proposals and other communication should be sent to your Publishing Editors at Springer.

Featuring compact volumes of 50 to 125 pages (approximately 20,000–45,000 words), Briefs are shorter than a conventional book but longer than a journal article. Thus, Briefs serve as timely, concise tools for students, researchers, and professionals.

Typical texts for publication might include:

- A snapshot review of the current state of a hot or emerging field
- A concise introduction to core concepts that students must understand in order to make independent contributions
- An extended research report giving more details and discussion than is possible in a conventional journal article
- A manual describing underlying principles and best practices for an experimental technique
- An essay exploring new ideas within physics, related philosophical issues, or broader topics such as science and society

Briefs allow authors to present their ideas and readers to absorb them with minimal time investment. Briefs will be published as part of Springer's eBook collection, with millions of users worldwide. In addition, they will be available, just like other books, for individual print and electronic purchase. Briefs are characterized by fast, global electronic dissemination, straightforward publishing agreements, easy-to-use manuscript preparation and formatting guidelines, and expedited production schedules. We aim for publication 8–12 weeks after acceptance.

Adrià Delhom · Alejandro Jiménez Cano ·
Francisco José Maldonado Torralba

Instabilities in Field Theory

A Primer with Applications in Modified
Gravity

 Springer

Adrià Delhom
Department of Astronomy and Astrophysics
Louisiana State University
Baton Rouge, LA, USA

Alejandro Jiménez Cano
Institute of Physics—Laboratory
of Theoretical Physics
University of Tartu
Tartu, Estonia

Francisco José Maldonado Torralba
Institute of Physics—Laboratory
of Theoretical Physics
University of Tartu
Tartu, Estonia

ISSN 2191-5423   ISSN 2191-5431 (electronic)
SpringerBriefs in Physics
ISBN 978-3-031-40432-0   ISBN 978-3-031-40433-7 (eBook)
https://doi.org/10.1007/978-3-031-40433-7

This Springer imprint is published by the registered company Springer Nature Switzerland AG
The registered company address is: Gewerbestrasse 11, 6330 Cham, Switzerland

Preface

Modified theories of gravity usually present new degrees of freedom, as well as higher order derivatives, wrong signs in certain terms and complicated couplings, either in the original Lagrangian or originated by the field redefinitions needed to reach an Einstein frame. As a consequence, they are very prone to present dynamical instabilities that could spoil any attempt to construct viable models within these frameworks. In this book, we introduce the most common types of instabilities that appear in field theory as well as some techniques to detect them, and supplement these contents with several examples. The goal is to understand the implications of having such behaviors and the application of these notions to modified theories of gravity.

This text is an extended and polished version of the lectures prepared for the course "Selected Topics in the Theories of Gravity", given at the Institute of Physics (University of Tartu, Estonia) in spring 2022.

The last chapter contains part of the results presented in recent peer-reviewed publications by the authors.

In order to follow the lectures smoothly, the reader is expected to have previous knowledge on differentials equations and classical field theory. For the last chapter, it is useful to have already some background in General Relativity theory.

For a better understanding, we included a list of exercises related to the topics introduced, and their solutions are provided at the end of the book.

Baton Rouge, USA   Adrià Delhom
Tartu, Estonia   Alejandro Jiménez Cano
Tartu, Estonia   Francisco José Maldonado Torralba
June 2023

Acknowledgements

The authors are deeply grateful to Jose Beltrán Jiménez for his mentoring on the topic, the discussions about it, and feedback on the manuscript. We also thank Gerardo García Moreno for useful comments and help with the quantum field theory discussion, and Verónica Errasti Díez for broadening our view on the discussed topics. This work has been done under the support of the Estonian Research Council and the European Regional Development Fund through the Center of Excellence TK133 "The Dark Side of the Universe". AJC was also supported by the Mobilitas Pluss post-doctoral grant MOBJD1035. FJMT is supported by the "Fundación Ramón Areces". AD is also supported by NSF grants PHY-1903799, PHY-2206557 and funds from the Hearne Institute for Theoretical Physics.

Contents

Acronyms

CECG   Cosmological Einsteinian Cubic Gravity
dof   Degree of freedom
EFT   Effective Field Theory
EoM   Equation of motion
FLRW   Friedmann-Lemaître-Robertson-Walker (metric)
GR   General Relativity
gRBG   (generalized) Ricci-based gravity theories
IR   Infrarred
KG   Klein-Gordon (theory)
MAG   Metric-Affine Gauge (gravity)
PDE   Partial differential equation
PG   Poincaré Gauge (gravity)
UV   Ultraviolet

Chapter 1
Introduction to Instabilities and Some Relevant Examples

Abstract In the first chapter we will provide some context for the readers to have a systematic understanding of what field theories are about, and how instabilities are defined within them. We will then go over examples of instabilities that are typically encountered in the literature, emphasizing their physical meaning. We will finally discuss how to interpret these instabilities within the effective field theory framework.

1.1 Introduction

One of the most relevant contributions from Newton was to provide a mathematical framework in which we can formulate problems involving time evolution of some physical quantity. Time evolution should be intuitively understood as the process in which the values of certain variables change, as time goes from one instant to the next one. Modern attempts to give a fundamental description of our universe are formulated as what is known as *field theories*. These are simply theories where the physical information is encoded in functions, called *fields*, over some base space. Depending on the formulation of the theory, the base space might be the spacetime, the phase space, or other sets with the required structure. This framework has been useful as well beyond the regime of fundamental physics, being able to provide accurate descriptions of more complex systems, finding important applications in more applied fields such as fluid dynamics, solid state physics, or any kind of system whose physical information can be encoded in a set of continuous variables.

From the mathematical perspective, in order to formulate a field theory, one first needs to provide a base space, in which the fields take values, and one usually expects it to have some nice mathematical properties that allow to introduce *derivatives*,[1] the basic mathematical objects needed to define evolution in a generalized sense. In this framework, now one can think of 'evolution' in space as well as in time, as one can think of the ways in which a mathematical function over spacetime changes from

---

[1] These properties are related to the topological and smooth structures of the base space, typically being a smooth manifold.

© The Author(s), under exclusive license to Springer Nature Switzerland AG 2023
A. Delhom et al., *Instabilities in Field Theory*, SpringerBriefs
in Physics, https://doi.org/10.1007/978-3-031-40433-7_1

point to point both in spatial and time directions. In fact, according to Relativity, there are no such things as absolute spatial and time directions. Hence, in order to describe time evolution in a field theory, one must first decide what direction in spacetime will be understood as time direction. This generally provides a splitting of spacetime into space + time in terms of a succession of hypersurfaces that cover the whole spacetime and do not intersect.[2] These hypersurfaces play the role of instants, and they can be used to define time evolution as the change undergone in physical quantities when passing from one of these surfaces to the next one. In general, when we use the word *evolution*, we will be referring to time evolution in this sense.

Having this base space, and the physically relevant functions over it to be described, the last ingredient to define the theory is something that establishes how the physical quantities evolve once we know their value at an instant of time (this is usually called *initial value problem*) or if we know their value at a suitable region of space (usually called *boundary value problem*). Partial differential equations (PDEs) are exactly the mathematical objects that we are looking for: they can provide a value for physical fields over the whole spacetime provided that we specify initial data or suitable boundary conditions. Thus, from a mathematical perspective, when dealing with field theories we are just dealing with a system of PDEs called *field equations* that describe a set of (physically relevant) functions over spacetime (or some other related space of interest). When one specifies the values at each point for one of these fields, we call that a particular *configuration* of the field. In modern theories, the field equations are usually obtained by assuming that the physical field configurations (i.e., the solutions to the field equations) are those which correspond to critical points of a certain functional of the fields that we call the *action*.

Once the field theory is specified, namely, once one has defined the set of fields to be described, the space they live on, and the PDEs that govern their behavior, the fundamental question to be answered is: what are the physically relevant configurations for these fields, and what are their properties? A general answer would be that any solution to the field equations is a candidate to be a physically viable configuration according to the theory. However, we can go further, and classify the properties of these solutions, regarding the type of initial/boundary data that gives rise to them, their asymptotic properties, and/or their compatibility with observed properties of the universe. Indeed, from a physical perspective, we know that we can never determine initial conditions exactly, so we might be wondering what happens if instead of having some initial data that determines an exact solution, we have initial data close to it. This can also be formulated by the question: what happens if we perform small perturbations of a given exact solution? Do the deviations with respect to the unperturbed solution remain arbitrarily small for sufficiently small perturbations, or does the perturbed evolution lead to a completely different physical system? In short, in the first case we say that such solution is stable under small perturbations, and in the second case we say that such solution is unstable under small perturbations (or perturbatively stable/unstable respectively). Thus, perturbative stability will be

---

[2] If you want to understand this more precisely, look up 'global hyperbolicity' and '3+1 decomposition'. Some recommendations of references are [59, 62].

characterized by assessing whether perturbations remain small through the whole evolution, or their corrections become comparable to the 'size' of the exact solution, thus pointing that these perturbations destabilize it.[3] This view gives a broad but imprecise idea of what instabilities are. The goal of these lectures is to provide a more precise description of these issues and, after that, to discuss their relevance in some modified gravity scenarios.

1.2   Warm Up Examples

In this section we aim to give a more precise definition of the basic types of instabilities arising in field theory but, first, let us provide with a little more context to the reader. In physics we aim to provide a mathematical description to observed physical systems. In such mathematical descriptions, it is mandatory to understand whether producing small modifications on them leads to small changes or completely destroys the system leading to other kind of phenomenology. For instance, imagine that one has a bound state of a planet orbiting a star, described by an exact solution to the two body problem. Now, one can decide to perturb this planet by, e.g., hitting it with a small asteroid. Will the orbit undergo a small change or, on the contrary, the planetary orbit will be destabilized sending it to infinity (or inside the star) and completely destroy the 2-particle bound state previously formed by planet-star? This corresponds to a physical example of the question *do the deviations with respect to the unperturbed solution remain arbitrarily small for sufficiently small perturbations, or does the perturbed evolution lead to a completely different physical system?* that we described, in more mathematical terms, in the previous section. Now, note that, if a system is strictly unstable, the probability that it forms and stays that way given some arbitrarily small set of initial conditions tends to zero. Therefore, one property that we require for the mathematical models that describe nature is that they are stable under perturbations which are small enough. Another question that might arise in the case of unstable systems is: how long does it take for the system to break apart? If the instability time is long enough, this system may be physically allowed only in some time window smaller than the instability time.

   To be clear, the question of stability refers to exact solutions of a theory, often called *vacua* or *backgrounds*, and it has to be studied in general case by case. Sometimes, when there are generic instabilities that arise for any vacuum of the theory, or when a background that is of physical relevance (e.g. Minkowski, de Sitter, etc.) is unstable, we say that the theory is unstable. But, strictly, stability is a property of exact solutions, not theories. Having said this, let us present some simple examples that will allow us to better understand how the analysis of these issues works, and to provide a classification of common types of instabilities arising in field theory:

---

[3] This can be extended to any observables related to that solution: If a small perturbation changes the value of an observable substantially after time evolution, then one must say that such observable is not stable under perturbations.

Example: Free scalar field around Minkowski

A free linear relativistic scalar field $\phi(t, \mathbf{x})$ is described by the real Klein-Gordon (KG) Lagrangian

$$\mathcal{L}_{KG} = \frac{1}{2} \partial^\mu \phi \partial_\mu \phi - \frac{1}{2} m^2 \phi^2 \tag{1.1}$$

which, upon extremizing the associated action functional, leads to the following field equations

$$\mathcal{E}_{KG}(\phi) \equiv \ddot{\phi} - \Delta\phi + m^2 \phi = 0. \tag{1.2}$$

Here the dots represent (partial) time derivatives, Δ is the standard Laplacian in $\mathbb{R}^3$ and m^2 is a constant. The general solution of this PDE can be expressed as follows in terms of Fourier modes,[4]

$$\phi_0(t, \mathbf{x}) = \int_{\mathbb{R}^3} d^3k \left[A_k \, e^{i(\mathbf{k}\cdot\mathbf{x} - \omega t)} + A_k^* \, e^{-i(\mathbf{k}\cdot\mathbf{x} - \omega t)} \right]_{\omega = \omega(\mathbf{k})}, \tag{1.3}$$

where

$$\omega(\mathbf{k}) := +\sqrt{|\mathbf{k}|^2 + m^2}, \tag{1.4}$$

and for arbitrary constants A_k. Provided that m^2 is real and non-negative, the integrand is an oscillatory (or constant) function which remains bounded over the spacetime. In other words, for each k, there is a real constant C_k such that the module of the integrand is always below C_k (actually any $C_k > \sqrt{2}|A_k|$).

The stress-energy tensor of the real free scalar field can be computed from the above Lagrangian. From there, one can obtain the energy density of a given field configuration for a given observer as the 00 component of the stress-energy tensor in coordinates $(t, \mathbf{x})$ adapted to such observer yielding

$$T_{00} = \frac{1}{2} \left[\dot{\phi}^2 + (\nabla\phi)^2 + m^2 \phi^2 \right]. \tag{1.5}$$

We see that this quantity is also non-negative provided that $m^2 \geq 0$. Note that this condition is crucial for having both oscillatory and positive-definite energy-density solutions to the field equations. Indeed, if $m^2 < 0$, then we might have exponentially growing/decaying solutions. While exponential decay does not destabilize a background, exponential growth is a typical cause of background instabilities due to unbounded growth of perturbations, as we will see later.

Having found exact solutions $\mathcal{E}_{KG}(\phi_0) = 0$, we can assess the stability of these solutions by studying how small perturbations of such backgrounds evolve. Consider an arbitrary (small) perturbation of such a solution, $\phi(t, \mathbf{x}) = \phi_0(t, \mathbf{x}) + \varphi(t, \mathbf{x})$; in this case, since the equation (in other words, the differential operator $\mathcal{E}_{KG}$) is linear,

---

[4] Note that we are assuming a folitation of Minkowski spacetime adapted to Cartesian coordinates.

$\mathcal{E}_{KG}(\phi) = \mathcal{E}_{KG}(\phi_0) + \mathcal{E}_{KG}(\varphi) = \mathcal{E}_{KG}(\varphi)$, so that from $\mathcal{E}_{KG}(\phi) = 0$ we obtain a field equation for the perturbations which is exact to all orders and is just the original Klein-Gordon equation, so that we already know the form of the solutions for the perturbations

$$\varphi_0(t, \mathbf{x}) = \int_{\mathbb{R}^3} d^3q \left[a_q e^{i(\mathbf{k} \cdot \mathbf{q} - \omega t)} + a_q^* e^{-i(\mathbf{q} \cdot \mathbf{x} - \omega t)} \right]_{\omega = \omega(\mathbf{q})} \tag{1.6}$$

Assuming that we perturb weakly each mode, namely that $|a_k / A_k| \ll 1$ at initial times, we have that $|\varphi_0 / \phi_0| \ll 1$ will be true at all times, guaranteeing the stability of the particular exact solutions ϕ_k as well as the general one, and therefore the validity of the perturbative expansion at all times. This is a trivial example of how to check whether a solution is stable: the key point is that the perturbative expansion is well defined at all times so that the deviations from the exact solution are controlled by some small parameter, preventing perturbations to grow to the 'size' of the original solution.

In the above example, the field equations for the perturbations can actually be derived to all orders due to linearity (wave superposition). This is a very particular feature of this example and will not arise in general cases. Indeed, in order to allow for interesting background solutions, one usually requires a system with various fields and/or nonlinearities to be present. In that case, the superposition principle will not apply, and one indeed has to compute the perturbation equations order by order. Let us warm up with a simple nonlinear example which allows to show this better:

Example: Scalar field with a cubic interaction

Note: Throughout this example we will omit the derivatives in the functional dependence (e.g., $\mathcal{L}(\phi) \equiv \mathcal{L}(\phi, \partial_\mu \phi)$) to alleviate the notation.
    Consider the following real scalar Lagrangian

$$\mathcal{L}(\phi) = \frac{1}{2} \partial^\mu \phi \partial_\mu \phi - \frac{1}{2} m^2 \phi^2 + \frac{\lambda}{2} \phi \partial^\mu \phi \partial_\mu \phi. \tag{1.7}$$

Its field equations read

$$\mathcal{E}(\phi) \equiv \Box \phi + \frac{\lambda}{2(1 + \lambda \phi)} \partial^\mu \phi \partial_\mu \phi + \frac{m^2}{1 + \lambda \phi} \phi = 0, \tag{1.8}$$

where $\Box$ is the d'Alembertian operator. These equations can have several branches of exact nontrivial solutions.[5] In particular, assuming $\partial_i \phi = 0$ (namely, homogeneity and isotropy), there exist nontrivial solutions $\phi_0(t)$. Now, we want to understand how perturbations behave around this background. To that end, consider $\phi(t, \mathbf{x}) = \phi_0(t) + \varphi(t, \mathbf{x})$. In full generality, we can expand the above field equations in φ, obtaining in general

$$\mathcal{E}(\phi) = \mathcal{E}(\phi_0) + \sum_{n=1}^{\infty} \mathcal{E}_{\phi_0}^{(n)}(\varphi), \tag{1.9}$$

where $\mathcal{E}_{\phi_0}^{(n)}(\varphi)$ is of order φ^n and depends on the background ϕ_0 around which the perturbative expansion is considered. Particularly, we obtain for linear perturbations the field equation

$$\mathcal{E}_{\phi_0}^{(1)}(\varphi) \equiv \ddot{\varphi} - \Delta \varphi + \frac{\lambda \dot{\phi}_0}{1 + \lambda \phi_0} \dot{\varphi} + \frac{2m^2 - \lambda^2 \dot{\phi}_0^2}{2(1 + \lambda \phi_0)^2} \varphi = 0. \tag{1.10}$$

We can see that linear perturbations $\varphi(t, \mathbf{x})$ for (1.7) around the vacuum ϕ_0 behave as wave like perturbations interacting with the background through a linear term in $\dot{\phi}$ and an effective mass term. The former will act as a damping/amplifying term depending on the sign of the background-dependent coefficient $(1 + \lambda \phi_0)^{-1} \lambda \dot{\phi}_0$. The sign of the effective mass term can also qualitatively change the behavior of the solutions as we will see later.

Note that the equations for linear perturbations can also be obtained from expanding the Lagrangian in perturbations as

$$\mathcal{L}(\phi_0 + \varphi) = \mathcal{L}(\phi_0) + \sum_{n=1}^{\infty} \mathcal{L}_{\phi_0}^{(n)}(\varphi), \tag{1.11}$$

where $\mathcal{L}_{\phi_0}^{(n)}(\varphi)$ is of order φ^n and depends on the background ϕ_0 around which the perturbative expansion is considered. Now, because field equations are obtained by varying the Lagrangian with respect to the corresponding field, in order to obtain the equations for perturbations up to order n we need to expand the Lagrangian up to order $(n + 1)$ in perturbations. Hence, in order to obtain the equations describing linear perturbations, we need to expand $\mathcal{L}(\phi_0 + \varphi)$ up to quadratic order in φ, finding

$$\mathcal{L}_{\phi_0}^{(1)}(\varphi) = (1 + \lambda \phi_0) \dot{\phi}_0 \dot{\varphi} + \left(\frac{\lambda}{2} \dot{\phi}_0^2 - m^2 \phi_0 \right) \varphi, \tag{1.12}$$

$$\mathcal{L}_{\phi_0}^{(2)}(\varphi) = \frac{1 + \lambda \phi_0}{2} \partial^\mu \varphi \partial_\mu \varphi + \lambda \dot{\phi}_0 \dot{\varphi} \varphi - \frac{m^2}{2} \varphi^2. \tag{1.13}$$

---

[5] You can check this using, e.g., Mathematica. For a reasonable computation time, try assuming $\partial_i \phi = 0$ (namely homogeneity and isotropy) and give some particular values to the constants. You will see that a complicated nontrivial solution exists.

Computing the field equations for the perturbations φ, you can check that $\mathcal{L}_{\phi_0}^{(1)}(\varphi)$ yields the equations for the background $\mathcal{E}(\phi_0) = 0$ which are satisfied by construction, and $\mathcal{L}_{\phi_0}^{(2)}(\varphi)$ yields the field equations describing linear perturbations $\mathcal{E}_{\phi_0}^{(1)}(\varphi) = 0$ if one writes $\ddot{\phi}_0$ in terms of ϕ_0 and $\dot{\phi}_0$ using the background equations. After integrating by parts, the Lagrangian (1.13) can also be schematically written as

$$\mathcal{L}_{\phi_0}^{(2)}(\varphi) = \frac{1}{2} G^{\mu\nu}(\phi_0) \partial_\mu \varphi \partial_\nu \varphi - \frac{\mu}{2} \varphi^2, \tag{1.14}$$

where $G^{\mu\nu}(\phi) = (1 + \lambda\phi)\eta^{\mu\nu}$ can be understood as an effective metric on top of which linear perturbations propagate and $\mu = \lambda\ddot{\phi}_0 + m^2$ can be understood as the squared effective mass of the perturbations. You might find other examples where $G^{\mu\nu}$ is not proportional to $\eta^{\mu\nu}$, which is usually referred as disformal couplings.

We can also consider examples where the background on top of which perturbations propagate is not a background of the perturbed field, but of other field in the theory:

Example: Massless scalar field in an homogeneous and isotropic universe

We consider an homogeneous and isotropic universe described by the Friedman-Lemaître-Robertson-Walker (FLRW) metric, which in suitable coordinates reads

$$g_{\text{FLRW}} = g_{\mu\nu} dx^\mu dx^\nu = dt^2 - a^2(t)\delta_{ij} dx^i dx^j, \tag{1.15}$$

and where $a(t)$ is the scale factor which describes the expansion of the universe. The Lagrangian describing a minimally coupled massless scalar field on this cosmological background is

$$\mathcal{L} = \frac{1}{2}\dot{\phi}^2 - \frac{1}{2a^2}(\nabla\phi)^2 = \frac{1}{2}g_{\text{FLRW}}^{\mu\nu}\partial_\mu\phi\partial_\nu\phi \tag{1.16}$$

which leads to the field equations

$$\mathcal{E}_{\text{FLRW}}(\phi) \equiv \ddot{\phi} + 3H\dot{\phi} - a^{-2}\Delta\phi = 0 \tag{1.17}$$

where $H := \dot{a}/a$ is the Hubble rate. Here a trivial background solution for the scalar field is a constant one, $\phi_0(t, \mathbf{x}) = C$, and the above Lagrangian and scalar field equations are appropriate for describing linear scalar perturbations as well. Note that the sign of the Hubble rate (namely expansion/contraction, as $a(t) > 0$) enters the equation as a damping/amplification term for the scalar perturbations. For the case where the perturbations get amplified because the universe is contracting, the scalar field will backreact on the metric through its coupling via the Einstein equations in a nontrivial way when the exponential growth becomes non-negligible.

1.3 Common Types of Instabilities and Their Physical Implications

The above examples are intended to give the reader a notion of what we mean by 'perturbations on top of a given background (or vacuum)'. The vacua can be exact solutions of the very same field that we are perturbing or exact solutions of other fields of the theory which couple to the fields that we want to perturb. Being familiar with this notions, we see that the result is some Lagrangian or field equations which, for the scalar field, typically can be written as modifications of the Klein-Gordon equation where the coefficients of each term depend nontrivially on the background.

To understand the basic types of instabilities that can arise, we can study the problem in a paradigmatic example, a quadratic scalar Lagrangian that describes linear scalar field perturbations in an isotropic background[6] without amplification/damping terms.[7] Our Lagrangian for scalar perturbations will thus be of the form

$$\mathcal{L} = \frac{1}{2}\left(a\dot{\varphi}^2 - b(\nabla\varphi)^2 - \mu\varphi^2\right), \tag{1.18}$$

where the coefficients a, b, μ encode relevant information about the background. Hence, these coefficients can in general depend on spacetime coordinates, but their variations will typically occur over time/length scales T, L that characterize the background where the scalar field is propagating on. We want to study the behavior of perturbations on much shorter time/length scales than the ones that characterize the background. The field equations are therefore

$$\mathcal{E}^{(0)}(\varphi) \equiv \ddot{\varphi} - \frac{b}{a}\Delta\varphi + \frac{\mu}{a}\varphi = \mathcal{O}(T^{-1}, L^{-1}). \tag{1.19}$$

Here, variations of the background-dependent coefficients, which are suppressed by the scales T and L, are sub-leading effects if the period/wavelength of the perturbations are kept small enough.[8] Since the background-dependent coefficients will be constant at leading order, we can try the following ansatz for the perturbations

$$\varphi_{\mathbf{k}} = A_{\mathbf{k}}e^{i\left(\sqrt{\frac{b}{a}}\mathbf{k}\cdot\mathbf{x} - \omega t\right)} + A_{\mathbf{k}}^* e^{-i\left(\sqrt{\frac{b}{a}}\mathbf{k}\cdot\mathbf{x} - \omega t\right)} \quad \text{where} \quad \omega = +\sqrt{\frac{b}{a}|\mathbf{k}|^2 + \frac{\mu}{a}} \tag{1.20}$$

To see how accurate it is, we can expand $\mathcal{E}^{(1)}(\varphi_{\mathbf{k}})$ in terms of (T^{-1}, L^{-1}), and see whether the proposed ansatz solves the leading order term. Indeed we find

---

[6] Although we could do the most general case without any symmetries, this case is enough to illustrate the most common types of instabilities, and isotropy facilitates the identification of the key aspects of the problem.

[7] These can also be relevant for stability, but they would obscure the analysis of the types of instabilities that we want to classify.

[8] Here the superindex (n) means that the expression is of order T^{-n}.

$\mathcal{E}^{(1)}(\varphi_{\mathbf{k}}) = \mathcal{O}(T^{-1}, L^{-1})$, since time/space derivatives of a, b, μ are of order T^{-1} and L^{-1} respectively, so that the proposed ansatz is a solution to leading order.

Now, note that the qualitative behavior of the solutions depend on the magnitude and combinations of signs of the background-dependent coefficients. While we have oscillatory solutions for $a > 0$, $b > 0$, $\mu \geq 0$, this is not true anymore for different sign combinations. For instance, if μ is negative and big enough, ω becomes imaginary, which will result in exponential growth of the $A_{\mathbf{k}}^{*}$ term. In this case, perturbations will grow unboundedly[9] and produce backreaction effects on the background such that the variation of the background cannot be neglected anymore (namely, the perturbative expansion in negative powers of (T, L) breaks down). When this happens, we say that such background is unstable under perturbations of the scalar.

In order to characterize the different qualitative behavior of the instabilities that can arise in this case, let us consider systematically the different sign combinations of the background-dependent coefficients a, b, and μ and study their dynamical implications. We will follow and extend the discussion in [58].

For illustrative purposes, it will suffice to consider a spatially homogeneous background slowly varying in time, though the arguments generalize in a straightforward manner. The resulting cases are:

- Stable case: $a > 0$, $b > 0$ and $\mu \geq 0$.
- Tachyonic instability: $a > 0$, $b > 0$ and $\mu < 0$.
- Gradient/Laplacian instability: $a > 0$, $b < 0$ or $a < 0$, $b > 0$.
- Ghost instability: $a < 0$ and $b < 0$.

1.3.1 Stable Case

This sign configuration leads to a wave equation with similar characteristics to the Klein-Gordon equation: it is a hyperbolic equation with oscillatory solutions whose frequency ω is always real, and whose energy density is always positive. Given their oscillatory behavior, it is straightforward to check that the perturbations of ϕ remain bounded at all times (and throughout space), and that they propagate at a speed given by $\sqrt{b/a}$ (in natural units). The causal character of the perturbations depends on the ratio b/a. If $b/a < 1$ the perturbations are subluminal speeds. If $b/a > 1$ the perturbations propagate at superluminal speeds.[10] Finally, for $b = a$ the perturbations travel at the speed of light. Although there is no problem with this, note that the smallest modification of the background (caused e.g. by the backreaction of the perturbations) could render $b > a$ and lead to superluminal propagation.

---

[9] There is no real constant which is bigger than $|\phi_{\mathbf{k}}|$ for all times in this case, no matter how small the initial amplitude of the perturbation is.

[10] Although this does not jeopardize the stability of the background, it is a sign that the corresponding field theory is not the low energy limit of a Lorentz invariant and unitary quantum theory [1].

1.3.2 Tachyonic Instability: Problems with Low-Momentum Modes

For this sign combination, the frequency of the modes with sufficiently small momenta becomes purely imaginary. This occurs for modes such that $|\mathbf{k}| < k_{\text{low}} := \sqrt{|\mu|/b}$, as the radicand defining ω becomes negative for such modes. Therefore, though high-momentum perturbations are still oscillatory, modes with momentum lower than k_{low} develop exponential growth or decay, since

$$\varphi_{\mathbf{k}} = A_{\mathbf{k}} e^{i\sqrt{\frac{b}{a}}\mathbf{k}\cdot\mathbf{x}} e^{|\omega|t} + A_{\mathbf{k}}^* e^{-i\sqrt{\frac{b}{a}}\mathbf{k}\cdot\mathbf{x}} e^{-|\omega|t} , \tag{1.21}$$

where we have substituted $\omega = i|\omega|$. While the exponentially decaying modes are perfectly stable, the problem lies with the modes that grow exponentially, also called *tachyonic* modes. Looking at (1.21), we see that though at early times $|\omega|t \ll 1$ the real exponentials are not relevant, at late times when $|\omega|t \gg 1$ the system is completely characterized by the exponential growth of the $e^{|\omega|t}$ modes, as the $e^{-|\omega|t}$ modes have decayed by that time. Therefore, we see that the would-be frequency of these modes does not have a physical meaning of 'how many times does this system undergo a periodic motion within a given time interval', but instead, it now signals the characteristic time at which exponential growth kicks in for each of the modes, which is given by $|\omega|^{-1}$. In a system in which tachyonic modes are excited, the characteristic time at which the instability kicks in $t_c = \omega_c^{-1}$ is the one corresponding to the tachyonic mode of fastest growth. In particular, if all tachyonic modes are excited, the characteristic time of the instability is related to the effective mass of the perturbations as $\omega_c = m_{\text{eff}} := \sqrt{|\mu|/a}$.

Due to the presence of two scales, one characterizing the changes in the background T and another characterizing the time at which the tachyonic instability becomes relevant, there could be different physical scenarios. In the case that $t_c \ll T$, the exponential growth of the tachyonic modes becomes dominant much before the background has shown any signs of evolution. Therefore, in this case, the background is unstable. Though the characteristic time of the background instability might depend on the exact way in which ϕ backreacts on it, it will generally be related to t_c through this dependence. In the opposite case, where $t_c \gg T$, the exponential growth of the low-momentum modes does not show up until the background has changed substantially. This situation allows to make physically sensible predictions related to the behavior of high-momentum modes in this background. Indeed, perturbation theory will provide physically meaningful predictions for all modes satisfying $\omega^{-1} \ll T$, since those modes will be insensitive to background evolution if the time intervals considered are short enough. Regarding the stability properties of the background, note that the tachyonic instability was derived assuming a constant background, and the background changes significantly much before the exponential growth of the tachyonic instability becomes relevant. Hence, in this case one cannot conclude that the background is unstable from an analysis where time-derivatives of the background of order $\mathcal{O}(T^{-n})$ have been neglected, and the problem must be

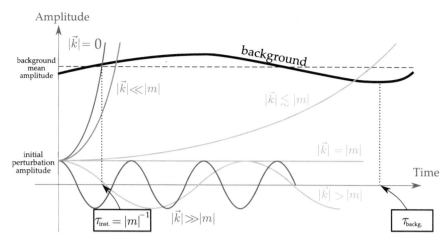

Fig. 1.1 Orientative plot of the modes of a system with a tachyonic instability characterized by $k_{low} = m$ and with with $T \gg t_c$. There is a fastest growing mode, namely the one with $|\mathbf{k}| = 0$, which grows as $\sim e^{mt}$. The perturbative approach can be employed for $t \ll m^{-1}$. [Reprinted from [40] with permissions, ©2021 A. Jiménez Cano]

studied taking into account the full dynamics of the background to truly assess its stability properties (Fig. 1.1).

As an example of a system where an instability is crucial to model correctly the ongoing physics, let us discuss the dynamics of a nearly homogeneous gas cloud with density $\bar{\rho}$ and pressure $\bar{P}$. Now, let us allow for the gas to have (arbitrarily) small density fluctuations, so that some regions are a little bit denser than others. The equations governing the system are conservation of matter and momentum (continuity equations) and Poisson's equation linking gravitational potential to mass density of the perfect fluid. Assuming a linear relation between density and pressure perturbations in the cloud $\delta P = c_s^2 \delta \rho$, density perturbations are described by

$$\partial_t^2 \delta \rho - c_s^2 \Delta \delta \rho - 4\pi G \bar{\rho} \delta \rho = 0 \qquad (1.22)$$

where G is Newton's constant. Here we can make the identification $a = 1$, $b = c_s^2$ and $\mu = -4\pi G \bar{\rho}$. Therefore we will have a tachyonic instability for modes satisfying $|\mathbf{k}| < k_J := \sqrt{4\pi G}/c_s$. This translates into an exponential growth of density fluctuations involving wavelengths larger than the Jeans' length $\lambda_J := 2\pi k_J^{-1}$. Hence, in a nearly homogeneous gas cloud, no matter how small are the density deviations from the average, if they occur over a large enough distance ($\gtrsim \lambda_J$), the density perturbations will grow quickly, implying a collapse of the cloud. Of course, this situation will in general stop at some point, when the fluctuations are big enough, fusion comes into play, changing the composition of the fluid and spoiling the linear relation between density and pressure perturbations. In that case the pressure of the collapsed fluid stabilizes the tachyonic growth leading to the formation of a star. We thus see that this instability is crucial to predict the formation of stars out of interstellar gas clouds and, indeed, we should view this instability as a transient regime from one unstable background $\bar{\rho} = $ constant (a homogeneous gas cloud) to another stable[11] background where matter has collapsed into a star possibly leaving some traces of the old gas orbiting around it. This example emphasizes the idea stated before that (tachyonic) instabilities are not always problematic, but they can be rather necessary to explain observed transient regimes in certain physical systems.

1.3.3   Gradient/Laplacian Instability: Problems with High-Momentum Modes

In this case we see that for UV modes ω becomes purely imaginary, thus triggering exponential growth/decay. Unlike the tachyonic case, for the gradient (also called Laplacian) instabilities the growth rate is arbitrarily fast, and therefore backgrounds supporting gradient instabilities are always unstable and therefore perturbation theory on top of them is physically meaningless. As a remark, note the in the case where $a\mu > 0$, modes satisfying $|\mathbf{k}| < \sqrt{|\mu/b|}$ do not develop instabilities (Fig. 1.2).

1.3.4   Ghostly Instabilities

Lastly, for negative a and b coefficients, we see that high-momentum modes are always stable, whatever the sign of μ, and for $\mu > 0$ a tachyonic instability will appear in the low-momentum regime. Thus, we see that, for a single field theory like (1.18), perturbations look stable for $\mu < 0$ and propagate a tachyonic instability for $\mu > 0$. Nevertheless, note that the associated Hamiltonian[12] is not bounded below (in the direction of increasing momenta), so that it does not have a ground state. Let us explore the consequences of this result both for classical and quantum theories.

Starting with classical theories, note that for the $\mu < 0$ case perturbations are oscillatory and for the $\mu > 0$ case there is a tachyonic instability. However, in both cases we have a Hamiltonian unbounded from below in the direction of increasing momentum. If the field is not interacting, conservation of energy will ensure that the system does not fall to such states.[13] In an interacting field theory, however, the field could decay to arbitrarily low (negative) energy states by exciting other

---

[11] As long as fusion is able to produce the necessary pressure to avoid further collapse!

[12] The Hamiltonian is T_{00} with $\dot{\phi}$ written as π/a, where π is the momentum conjugated to ϕ.

[13] Note that the background is time dependent, so energy conservation is only approximate. However, the breaking terms will be of order $O(T^{-1})$, so whenever they become relevant, our analysis of stability is no longer valid. Therefore energy is conserved to the precision to which our analysis is valid. As well, there might be other conserved quantities that delimit the phase space region that is dynamically reachable from given initial conditions. If such region does not include arbitrarily high momentum states, then the energy of the system will also be dynamically bounded.

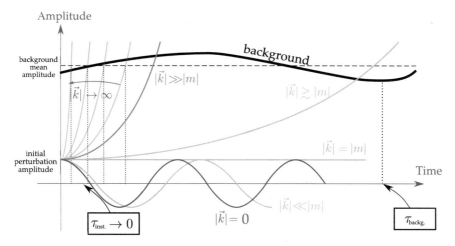

Fig. 1.2 Orientative plot of the modes of a system with a gradient instability (for the case $\mu/b > 0$). There are arbitrarily fast growing modes that spoil the perturbative approach at arbitrarily short times. [Reprinted from [40] with permissions, ©2021 A. Jiménez Cano]

fields to arbitrarily high-energy states in what is usually called a *ghost* (or runaway) instability. Though the energy of the whole system is conserved, this can lead to strange behaviors. For instance, considering a gas of interacting particles through elastic scattering, by having a particle with such a runaway instability, we can see that though the energy of the gas will be kept constant, the temperature and pressure of the gas will grow exponentially with time due to the interactions with the runaway particle, which increase the mean velocity of the particles in the gas. Of course, for this type of instability to become manifest, the runaway particle should be allowed to access regions of phase space with arbitrarily high momentum. Was there any set of symmetries that forbids this region by enforcing conservation laws, the dynamics of such system would avoid this type of instability.

Even if this lectures deal with classical theory, a comment regarding ghosts at a quantum level is in order, since typically we regard classical theory as an approximation to quantum. At the quantum level ghosts are problematic, so even if we managed to avoid exciting the ghost classically, at the quantum level it will make the vacuum to spontaneously decay. If a ghostly field φ is not interacting there is no problem due to conservation of energy. However, if it is interacting e.g. with another field χ with a term of the form $\varphi^2 \chi^2$, it renders any state $|0\rangle$, defined as the vacuum of the theory, unstable. Indeed, since $|0\rangle$ is not a ground state of the Hamiltonian (it is unbounded from below), the process[14]

$$|0\rangle \rightarrow \varphi + \varphi + \chi + \chi \tag{1.23}$$

---

[14] Note that in any realistic theory φ will interact with any other field at least gravitationally due to the universality of the gravitational interaction, and this process can be mediated at least by graviton exchange.

would be allowed by conservation of energy for arbitrarily high momentum of the outgoing particles. The decay rate of this process involves a divergent phase space integral (see e.g. [18, 33]), so the decay of the vacuum is instantaneous. See Sect. 1.6 for the reinterpretation of this in the context of effective field theories.

1.4 Generalization for N Degrees of Freedom

Let us now generalize the identification of different types of instabilities explored above to a theory with several degrees of freedom (dof(s) for short). As you might have noticed at this point, the presence or absence of instabilities for any dof of the theory has to do exclusively with the coefficients that define the differential operator which controls their dynamics. If we have N propagating[15] dofs, no matter how they transform under different transformation groups, we can always build an N-tuple ϕ^I out of them with $I = 1, ..., N$. By doing so, we will be able to express the linear part of any Lagrangian describing the propagation of N dofs on top of an isotropic background as

$$\mathcal{L} = \frac{1}{2}\left(a_{IJ}\dot{\varphi}^I\dot{\varphi}^J - b_{IJ}\partial^i\varphi^I\partial_i\varphi^J - \mu_{IJ}\varphi^I\varphi^J\right), \tag{1.24}$$

independently of how each of the components of ϕ^I transforms under different transformation groups. The field equations derived from the above Lagrangian are

$$a_{(IJ)}\ddot{\varphi}^J - b_{(IJ)}\Delta\varphi^J - \mu_{(IJ)}\varphi^J = \mathcal{O}(T^{-1}, L^{-1}). \tag{1.25}$$

The matrices a_{IJ} and μ_{IJ} are called kinetic and mass matrix, and we will call b_{IJ} the gradient matrix.[16] They determine the exact differential operator defining each of the PDEs that governs the propagation of each dof. Though redundant, the emphasis on propagating was put because the term dof is sometimes used in the literature to call any field appearing in the Lagrangian. However, if the Lagrangian is that of a constrained theory, some of the fields appearing will not propagate. In order to carry on this analysis, one must first solve the constraints and write the Lagrangian only in terms of truly propagating fields. For example, in a Proca theory, usually formulated in terms of four fields A_μ that transform as a vector under Lorentz transformations, A_0 is non-dynamical, and there is a constraint equation yielding $A_0 = \partial_i \dot{A}^i$. One needs to integrate out the non-dynamical A_0 field by using such constraint in the Lagrangian before carrying the stability analysis as outlined here.

---

[15] Here we want to emphasize the word propagating, for reasons explained below.

[16] We use the usual notation for total (anti)symmetrization of an object with indices where $2A_{[ij]} = A_{ij} - A_{ji}$, $2A_{(ij)} = A_{ij} + A_{ji}$, and the same applies for n indices with the appropriate $n!$ factor on the left-hand side.

To assess stability of the system defined by (1.24), we just have to look at the positiveness properties of the kinetic, gradient and mass matrices. Negative eigenvalues will be tied to instabilities in the same way than for the single-field case. Particularly, if the different matrices commute, they can be diagonalized together, and one can look at the stability properties of each field-eigenvector by analogy to the single-field case. On the other hand, performing the stability analysis in a case in which they cannot be simultaneously diagonalized (or diagonalized at all) requires a detailed study of the particular system.

As a remark, let us discuss what are these field-eigenvectors associated to the different matrices and how they are related to the original fields. To do that, we should introduce the concept of *field redefinition*. In the same way, we can do coordinate transformations in classical point-particle mechanics in order to work with any desired set of phase space coordinates, we can also do this kind of transformations in field theory, and work with different field variables. Different field variables will generally have different physical meaning, but all will encode the same information about the system if the transformation from one to the other is well defined.[17] Having said this, note that if an eigenvector of one of these matrices is α, with components α_I in the basis in which the matrix is written, it will define a linear combination of the original fields by $\Psi_\alpha = \alpha_I \phi^I$, with each eigenvector yielding a different linear combination of the original fields. Thus these field-eigenvectors are nothing but a linear field redefinition of the original fields which takes the corresponding matrix into diagonal form. Using these field space variables, if they exist, the question of what dofs of the system are unstable and what kind of instability do they develop becomes straightforward by analogy with the single-field case.

1.5 Strongly Coupled Backgrounds and Their Unstable Nature

Another type of instability that often arises in field theories is the so-called *strong coupling* instability. Unlike the ones discussed above, this type of instability cannot be explicitly seen at linear order in perturbation theory, although one might already suspect its presence from the linear analysis. This is so because this instability occurs when some dof of a theory does not propagate on top of a particular vacuum because the kinetic term of such mode vanishes on such background. So, in short, one would expect this kind of instability to occur when there is a particular background where some of the eigenvalues of the kinetic matrix vanish, but they do not vanish in some field configurations that are arbitrarily close to such background. Thus, naively, if we canonically normalize the kinetic term of the propagating dofs,[18] we expect that the coefficients of the rest of the terms in the Lagrangian diverge.

---

[17] Mainly invertible and with some required number of invertible derivatives.

[18] Think of it roughly as multiplying (1.25) by $(a_{(IJ)})^{-1}$ on the left.

To see this, consider a theory depending on a set of fields $\{\phi^I\}$, and a background $\{\Phi_0^I\}$ around which the perturbations of one of the fields (that we will call ψ) do not propagate. Consider also that we approach such a background with a curve in solution space $\{\phi_0^I(\lambda)\}$ with $\lambda \in [0, 1]$, such that for $\lambda < 1$ there is no strong coupling (i.e., ψ propagates) and $\{\phi_0^I(1) = \Phi_0^I\}$. If we expand around any point of this curve with $\lambda < 1$, we can extract from the Lagrangian the kinetic term of the perturbations ψ and it will have the form:

$$\mathcal{L}_\lambda = \frac{1}{2} F(\phi_0^I(\lambda)) \dot{\psi}^2 + \mathcal{L}_{\text{rest}} \,. \tag{1.26}$$

This Lagrangian must satisfy (by construction) that $F(\phi_0^I(1)) = F(\Phi_0^I) = 0$. Notice that if we canonically normalize the perturbation,

$$\mathcal{L}_\lambda|_{\text{canon.}} = \frac{1}{2} \dot{\psi}^2 + \frac{1}{F(\phi_0^I(\lambda))} \mathcal{L}_{\text{rest}} \,, \tag{1.27}$$

all the interactions (non-kinetic couplings) of ψ will become infinite around $\{\Phi_0^I\}$

$$\lim_{\lambda \to 1} \frac{1}{F(\phi_0^I(\lambda))} \mathcal{L}_{\text{rest}} = \infty \,. \tag{1.28}$$

This is the reason why it is said to be strongly coupled.

In general, we expect this to happen when we have a theory that propagates N dofs in the general case but linear perturbations on top of a particular vacuum of the theory describe $M < N$ propagating modes. In that case, it may (or may not) occur that we are able to see explicitly how the strongly coupled modes are present by going at higher-order in perturbations. However, there is no general rule to know at which order will those appear. This discontinuity in the number of dofs when going to higher-order perturbations signals the breakdown of the perturbative expansion of the solution (see Sect. 3.3), which then yields meaningless results on top of a strongly coupled background. This occurs because the system of differential equations describing time evolution have an abrupt change in their time-derivative order. From the phase space perspective, strongly coupled backgrounds can typically be linked to some geometrical singularity in phase space. Note that, though this instability can be guessed when one knows the number of propagating dofs of a theory in the general case, it is very hard to foresee in cases where this information is not available since, in that situation, one does not know whether perturbations around a particular background describe less dofs than around other backgrounds. Thus, this instability will generally stay invisible unless we come across the situation where we know the number of propagating dofs on top of two different backgrounds does not coincide, or we find a change in the number of dofs at different orders in perturbation theory.

1.6 Instabilities and Effective Field Theories

(This section can be skipped on a first read)

In this section, we briefly comment on the re-interpretation of these problems within the effective field theory (EFT) framework. Broadly speaking, an EFT is a quantum field theory which is valid (well) below a given energy scale Λ that we call the cutoff of the EFT. An EFT can be obtained as a low energy description of a high-energy theory after performing the path integral over the dofs with masses above the cutoff scale. These dofs that have been *integrated-out* are called ultraviolet (UV) dofs, and the high-energy theory from which the EFT is derived is called *UV-completion* of the EFT. In this case, the cutoff of the EFT is usually of the order of the mass of the lightest dof that has been integrated-out. The UV dof would only be excited when we reach energies of the order of the cutoff.[19]

The idea behind the EFT framework is that we should be able to explain all physical phenomena related to processes with energies below the cutoff only in terms of the infrared (IR) or low-energy dofs, which are the actual dofs of the EFT (in particular their masses must be below the cutoff). This is based on the fact that observables admit an asymptotic expansion on the energy of the process E over the masses of the dofs involved on it, i.e., an expansion in E/Λ. However, the EFT is not completely blind to the UV physics from which it is derived: the process of integrating out UV dofs generates a tower of interactions with higher mass-dimension operators (a.k.a. irrelevant operators) suppressed by powers of the cutoff.

Even though EFTs can be obtained from these low-energy asymptotic expansions of UV theories, one does not need a UV theory to formulate an EFT. In fact, the power of the framework relies precisely on this fact. Given a set of dofs which we want to describe, we can ask what is the most general Lagrangian that can be written to describe them that is compatible with a given set of symmetries. This is indeed a Lagrangian with an infinite number of operators of ever-growing mass-dimension that have to be suppressed[20] by powers of some energy scale Λ, above which the theory fails to be perturbatively unitary. This kind of Lagrangian will be able to describe the low-energy phenomena associated to those IR dofs provided that the symmetries that have been enforced are symmetries of the (unknown) UV theory, regardless of how many dof the UV theory has, and of what kind they are. As famously repeated by many high-energy physicists, from this perspective, EFTs allow us to parametrize our ignorance in a very systematic way, and can be used to unveil UV physics by precision experiments. Indeed, strictly speaking, since we will never be sure of whether a theory that can be regarded as a fundamental theory is indeed the correct description of the universe, we should always understand any theory as an EFT, perhaps in some generalized sense.

---

[19] Since we are dealing with (healthy) quantum dofs, an excitation of such dofs is a quanta (or particle) with a given energy, which cannot be smaller than its mass.

[20] Recall that the mass dimension of a Lagrangian density is the number of spacetime dimensions.

After this digression on EFTs, let us come back to our topic. As we have explained, no matter how we obtain an EFT, if as a low-energy limit of some known theory or as a parametrization of our ignorance of the high-energy physics, we will always end up with a Lagrangian with higher-order time derivatives coming from the irrelevant operators of the EFT. As we will see in Lecture 2, these higher-order terms generically introduce ghost dofs due to Ostrogradski theorem. However, typically, these unstable dofs are only excited above the cutoff scale of the theory, and therefore are not relevant to analyze the behavior of the EFT.

1.6.1 Strong Coupling

The strong coupling is only a problem if it is regarded as a complete theory. If we treat the higher-order terms of the EFT (the ones that would-be related to UV dofs) as perturbative corrections, everything works fine as long as we do not explore energy scales close or above the cutoff. This is not surprising, but a feature of EFTs, which contain higher-order derivative terms (this is a reminiscent of the dofs that have been integrated out). If we want to approach such scales, then one needs to take into account processes that will depend on what UV completion of the EFT is employed. See the recent papers [17, 26, 39], where this issue is discussed in the context of the strong coupling problem of cubic theories of gravity. We will come back to this in Sect. 3.3.

As a remark, note that any background with typical variation scales corresponding to energies above the cutoff will be able to excite the would-be dofs corresponding to the higher-order derivative operators. These are not dofs of the EFT and, the EFT is not well defined when these dofs are excited. Therefore, around such a background, the EFT will not yield any physical predictions.

1.6.2 Ghost and Gradient Instabilities

Regarding ghosts, again, we have to separate two cases, when they are only excited above the EFT cutoff, and when they are IR dofs, so that they get excited within the regime of validity of the EFT. In the first case, they are not part of the spectrum of the EFT, they are artifacts of extrapolating our theory beyond its regime of validity, and therefore we cannot draw any conclusion from them. In the second case, which we call IR ghosts, and they are not always catastrophic due to the cutoff of the theory Λ acting as a regulator for the decay rate of the vacuum. This is the case because regions of phase space with energies above the cutoff are outside of the regime of validity of the EFT and, therefore, any computation that relies on the EFT but involves those regions will not be physically meaningful. In practice, this restriction delimits the available volume of phase space which is available for the process to occur, bringing the decay rate of the vacuum to a finite value proportional to such a volume. We can be a bit more specific by considering the following case. The rate at which a tree-level process mediated by an interaction with a coupling α with mass dimension d occurs will typically be of order $\alpha^2 \Lambda^{4+2d}$. If α is a *natural* $\mathcal{O}(1)$ coupling, then the decay rate of the vacuum becomes of the order of the fastest processes described by the EFT (faster processes excite modes above the cutoff), usually rendering the theory useless for predictions. However, in some cases, α can be very small so the timescale for the decay of the vacuum can be much slower than some processes available within the EFT. In that case, it is possible to infer an upper limit for the cutoff scale of any EFT containing IR ghostly dofs. An example of such *unnatural* coupling would be a ghost that does not couple to anything in the Standard Model except through gravity (recall that gravity couples universally). In that case, by considering the process $|0\rangle \to \varphi + \varphi + \gamma + \gamma$ mediated by gravity (with φ a ghost, $\alpha = M_P^{-2}$, and γ representing photons), the decay rate of the vacuum will be of order $M_P^{-4} \Lambda^8$. From the observations of the spectrum of gamma rays coming from outer space, and one can obtain the upper bound $\Lambda \lesssim 3 \, \text{MeV}$ [21].

Let us now comment on the case of gradient instabilities. If the unstable modes are active below the cutoff Λ, the instability will also have a time scale associated to it $\sim \Lambda^{-1}$. Hence, we have to distinguish two cases: for $a\mu \leq 0$ the IR modes are still unstable, and the time scale of the instability will be related to the cutoff. For $a\mu > 0$, modes of sufficient low momenta are stable. If the cutoff is below the threshold above which the gradient instability develops ($|\mathbf{k}| < \sqrt{|\mu/b|}$), no unstable modes will actually be present within the regime of validity of the EFT. In this case, the instability is an artifact of trusting the EFT beyond its regime of validity, so the theory is physically meaningful.

As a final remark, let us point out that some authors have also attempted to provide quantization schemes in which the presence of ghostly dofs does not lead to physical inconsistencies. This is done by relaxing the usual hermiticity requirement for the Hamiltonian, so that only antilinearity is required. This is enough to have real eigenvalues, but its applicability as a realistic theory is still a matter of debate among the community [13, 19, 61]. Also the possibility that theories with ghosts are actually stable due to nonperturbative mechanisms has been discussed [5]. Nonperturbative solutions to non-linear theories that play this role are called ghost condensates.

Chapter 2
Ostrogradski Theorem and Ghosts

Abstract In the second chapter we will go over the celebrated Ostrogradski theorem that applies to theories with higher-order time derivatives. We will use a formalism that treats point particle classical mechanics and field theories on an equal footing. To understand the theorem, we will see how to define a Hamiltonian for higher-order theories, and how these are always unbounded in phase space, concretely in the direction where the momenta associated to the higher-order degrees of freedom grow. We will also show how the ghost of a vector relativistic theory with a non-gauge invariant kinetic term can be reformulated as an Ostrogradski ghost.

2.1 Some Preliminary Ideas

The Lagrangians we considered in the previous chapter contain, as usual, at most first-order derivatives of the fields. Here, we will focus on the cases containing higher-than-first-order derivatives. Usually, these derivatives imply the need of additional initial conditions to determine the evolution of the system, showing that the fields of the theory hide extra dofs (which can be extracted with appropriate field redefinitions). As we will see, when these extra dofs are present, ghosts inevitably appear and generically propagate around the solutions of the theory, making them invalid for physical purposes. To that end, we will resort to the Hamiltonian formalism, starting by quickly revisiting the 'standard' Hamiltonian analysis for field theories, and then extending it to theories with higher-than-first-order derivatives.

Consider a theory (in flat space) depending on some family of fields ϕ^I and up to first-order derivatives of them,

$$S[\phi^I] = \int \mathcal{L}(\phi^I, \partial_i \phi^I, \dot{\phi}^I) \mathrm{d}^4 x = \int L[\phi^I, \dot{\phi}^I] \mathrm{d}t , \qquad (2.1)$$

where we are denoting $\dot{X} := \partial_0 X$ and the indices i, j... cover the spatial directions. Here we have introduced the Lagrangian functional $L[\phi^I, \dot{\phi}^I]$,

A. Delhom et al., *Instabilities in Field Theory*, SpringerBriefs
in Physics, https://doi.org/10.1007/978-3-031-40433-7_2

$$L[\phi^I, \dot\phi^I] := \int \mathcal{L}(\phi^I, \partial_i\phi^I, \dot\phi^I) \mathrm{d}^3 x \,. \tag{2.2}$$

The Euler-Lagrange equations are

$$0 = \frac{\delta S}{\delta\phi^I} = \frac{\delta L}{\delta\phi^I} - \frac{\mathrm{d}}{\mathrm{d}t}\frac{\delta L}{\delta\dot\phi^I} = \frac{\partial \mathcal{L}}{\partial\phi^I} - \partial_\mu\frac{\partial \mathcal{L}}{\partial(\partial_\mu\phi^I)}, \tag{2.3}$$

and, after applying the chain rule in the last term, we get

$$0 = (\text{independent of } \{\ddot\phi^J\})_I - \mathcal{K}_{IJ}\ddot\phi^J \,. \tag{2.4}$$

Here, we defined

$$\mathcal{K}_{IJ} := \frac{\partial^2 \mathcal{L}}{\partial\dot\phi^I \partial\dot\phi^J}, \tag{2.5}$$

called the *Hessian matrix*. For the Lagrangians (2.1) that we normally use in physics, the Hessian corresponds with the kinetic matrix (maybe up to normalization). In addition, we assume that the theory fulfills the *non-degeneracy condition*,[1]

$$\det(\mathcal{K}_{IJ}) = \det\left(\frac{\partial^2 \mathcal{L}}{\partial\dot\phi^I \partial\dot\phi^J}\right) \neq 0 \,, \tag{2.6}$$

because this allows to re-express the Euler-Lagrange equations as

$$\ddot\phi^J = (\mathcal{K}^{-1})^{JI} \times (\text{independent of } \{\ddot\phi^J\})_I \,. \tag{2.7}$$

In other words, we are able to solve for the accelerations in terms of the fields and the velocities:

$$\phi^I(t, \mathbf{x}) = f^I(t, u(\mathbf{x}), \dot u(\mathbf{x})) \,, \tag{2.8}$$

where u and $\dot u$ represent the initial conditions for the field and its velocity, respectively, in a spacelike hypersurface.

In flat space, for this theories, we can perform the Hamiltonian construction as follows. First introduce the canonical variables:

$$q^I := \phi^I \,, \qquad \pi_I := \frac{\delta L}{\delta\dot\phi^I} = \frac{\partial \mathcal{L}}{\partial\dot\phi^I} \,. \tag{2.9}$$

---

[1] We will give a more general definition for 'Hessian' and 'non-degeneracy' later, which will be valid for also for Lagrangians with higher-than-first-order derivatives.

If the Lagrangian is non-degenerate we can solve for the velocities in the definition of the momenta, $\dot{q}^I = \dot{q}^I(\pi_J)$, and we are able to introduce the Hamiltonian functional:[2]

$$H[q^I, \pi_I] := \int \dot{q}^I|_{q,\pi} \, \pi_I \, d^3x - L|_{q,\pi} = \int \left(\dot{q}^I|_{q,\pi} \, \pi_I - \mathcal{L}|_{q,\pi} \right) d^3x , \qquad (2.10)$$

where $X|_{q,\pi}$ indicates that X should be written in terms of the canonical variables. The dynamics is then determined by the Hamilton equations:

$$\dot{q}^I = \frac{\delta H}{\delta \pi_I}, \qquad \dot{\pi}_I = -\frac{\delta H}{\delta \phi^I} . \qquad (2.11)$$

These equations require initial conditions for q^I and for π_I, which are essentially equivalent to those of ϕ^I and $\dot{\phi}^I$ in Lagrangian formulation. It is worth remarking that a *true* (or *propagating*) dof requires is a pair of initial conditions, one for itself and another one for its velocity or momentum.

In the next section we are going to extend this procedures to Lagrangians with higher-than-first-order derivatives. As we will see, in such a case, a single field ϕ could lead to two (or more) dofs.

2.2 Lagrangian with Higher-Order Derivatives

It is clear, by construction, that the formalism of the previous section is not valid for Lagrangians containing higher-than-first-order derivatives (except when they can be eliminated after integration by parts). The generalization of the previous formalism was done in 1850 by Ostrogradski in his work [53]. To deal with it, we first introduce the following concepts.

Definition 2.1 (*Hessian*) The *Hessian of a Lagrangian L* is the matrix of second variations of L with respect to the highest time derivatives of the fields (after removing the global Dirac delta).

The previous definition might seem a bit obscure,[3] but for the cases we are going to deal with in these lectures, the Hessian can be written in a more familiar way as the matrix of second partial derivatives of $\mathcal{L}$ with respect to the highest time derivatives of the fields.

---

[2] Notice that if the condition (2.6) is not fulfilled, then the velocity cannot be solved in terms of the momentum. This is signaling the presence of constraints that should be appropriately treated (e.g., via the Dirac procedure).

[3] The issue of the delta is due to the fact that second variations of integral functionals are proportional to the Dirac delta or spatial derivatives of it. So, after computing the matrix of second variations, one should multiply by a test function and integrate to remove the delta. The result is a matrix (in the space of field variables) whose entries are linear operators. Such a matrix is the Hessian.

Remark If we introduce new fields and constraints (via Lagrange multipliers) so that only first-order time derivatives appear (the so-called *first-order reduction* of the theory), the Hessian coincides with the kinetic matrix of the reduced Lagrangian. This is the approach followed for example in [30, 43], and allows for a more systematic method to study the dofs of theories with higher-order derivatives. Nonetheless, for our purposes it is more convenient to work with the Lagrangian in its higher-order form.

Definition 2.2 (*Non-degenerate Lagrangians*) A Lagrangian is said to be *(non-)degenerate* if the Hessian has (non-)vanishing determinant.

Degeneracy is connected with the fact that some of the highest-order derivatives can be eliminated for instance after integration by parts or by redefining the fields appropriately. Let us see a very simple example of a degenerate with first derivatives:

Example: Trivial degenerate Lagrangian

For instance, consider the theory,

$$S[\phi] = \int (\phi\dot{\phi} - \phi^2)\mathrm{d}^4 x. \tag{2.12}$$

It only depends up to first-order derivatives of one field, so we do not expect more than one dof.

Let us first check that it is degenerate:

$$\frac{\delta L}{\delta\dot{\phi}} = \frac{\partial L}{\partial\dot{\phi}} = \phi \quad \Rightarrow \quad \frac{\delta^2 L}{\delta\dot{\phi}^2} = 0. \tag{2.13}$$

This implies that we cannot solve for the field in terms of 'position+velocity' initial conditions (as it should be for one dof)

$$\phi(t, \mathbf{x}) \neq f(t, u(\mathbf{x}), \dot{u}(\mathbf{x})). \tag{2.14}$$

In fact, one can extract appropriate boundary terms in the time direction to eliminate them:

$$S[\phi] = \int \left[\partial_0 \left(\frac{1}{2}\phi^2 \right) - \phi^2 \right] \mathrm{d}^4 x = \int (-\phi^2)\mathrm{d}^4 x + \text{boundary term} \tag{2.15}$$

So this theory is clearly non-dynamical. The only solution is $\phi = 0$ which has no non-trivial time evolution. So this theory propagates no dofs.

Indeed, if we go to the Hamiltonian picture (we call $q := \phi$) the degeneracy leads to a constraint when computing the momenta (primary constraint):

$$\left(\frac{\delta L}{\delta\dot{\phi}} =: \right) \pi = q \quad \Rightarrow \quad C := \int \mathrm{d}^3 x (\pi - q). \tag{2.16}$$

The Hamiltonian of the theory is

$$H[q, \pi] = -\int q^2 \mathrm{d}^3 x + \lambda C(q, \pi) \,, \qquad (2.17)$$

where λ is a Lagrange multiplier. The time evolution of the constraint C leads to another constraint $\mathcal{D} = q$, which finishes the Dirac algorithm. It can be checked that the Poisson bracket of these two constraints is a constant number. Therefore, we have two second-class constraints, and this reduces in one unit the number of dofs. So, in conclusion, the theory propagates 0 dofs (i.e., there are no dynamical variables in it), as we showed previously in the Lagrangian formulation.

Consider now a field theory depending on just one scalar field and up to second-order time derivatives:

$$L[\phi, \dot\phi, \ddot\phi] = \int \mathcal{L}(\phi, \partial_i \phi, \dot\phi, \partial_i \partial_j \phi, \partial_i \dot\phi, \ddot\phi) \, \mathrm{d}^3 x \,. \qquad (2.18)$$

As we already said, this Lagrangian functional is assumed to be non-degenerate. In this case, this is true if and only if

$$\frac{\delta^2 L}{\delta \ddot\phi^2} \neq 0, \qquad (2.19)$$

which in this case is equivalent to[4]

$$\frac{\partial^2 \mathcal{L}}{\partial \ddot\phi^2} \neq 0 \,. \qquad (2.20)$$

Similarly, as in the previous section, now the Euler-Lagrange equation

$$0 = \frac{\delta S}{\delta \phi} = \frac{\delta L}{\delta \phi} - \frac{\mathrm{d}}{\mathrm{d}t} \frac{\delta L}{\delta \dot\phi} + \frac{\mathrm{d}^2}{\mathrm{d}t^2} \frac{\delta L}{\delta \ddot\phi} = \frac{\partial \mathcal{L}}{\partial \phi} - \partial_\mu \frac{\partial \mathcal{L}}{\partial (\partial_\mu \phi)} + \partial_\mu \partial_\nu \frac{\partial \mathcal{L}}{\partial (\partial_\mu \partial_\nu \phi)} \qquad (2.21)$$

can be written, after using the chain rule, as

$$\overset{....}{\phi} = \left(\frac{\partial^2 \mathcal{L}}{\partial \ddot\phi^2} \right)^{-1} \times (\text{something independent of } \overset{....}{\phi}) \,, \qquad (2.22)$$

---

[4] The second variations of L are distributional objects proportional to the Dirac delta or spatial derivatives of it. In the cases we will study in these lectures no derivatives of the delta appear in the Hessian. In these cases, one can multiply by a test function and integrate to eliminate the delta, and the result is the matrix of second partial derivatives of $\mathcal{L}$ (see e.g. (2.65)).

which locally admits solutions of the type

$$\phi(t, \mathbf{x}) = f\left(t, \ u(\mathbf{x}), \ \dot{u}(\mathbf{x}), \ \ddot{u}(\mathbf{x}), \ \dddot{u}(\mathbf{x})\right). \tag{2.23}$$

As in the previous section, $\{u, \dot{u}, \ddot{u}, \dddot{u}\}$ are the initial conditions for $\{\phi, \dot{\phi}, \ddot{\phi}, \dddot{\phi}\}$, respectively, in a spacelike hypersurface. Observe how the presence of one additional derivative order in the Lagrangian requires two additional initial conditions, $\ddot{u}$ and $\dddot{u}$. Each pair of initial conditions correspond to a propagating dof, so here we have an example of a scalar field that contains two dofs. The extra one is hidden in the fact that there are higher-than-first-order derivatives in the (non-degenerate) Lagrangian.

Let us remark one more time that the non-degeneracy condition is what allows us to solve for the highest-order derivative terms of the equations of motion, which are of the order $2\times$ the highest order of the Lagrangian. For instance, in the previous case, in which the Lagrangian depends up to second derivatives of ϕ, the non-degeneracy permits to solve the equations for $\dddot{\phi}$. Now we present an example of a particular Lagrangian depending on two fields, showing how all of these ideas can be straightforwardly generalized to multi-field theories:

Example: Lagrangian with two fields up to 2nd-order derivatives of one field

Consider the Lagrangian

$$\mathcal{L} = \frac{1}{2}\ddot{\phi}^2 + \frac{1}{2}\dot{\psi}^2 + a\ddot{\phi}\dot{\psi} + \frac{1}{2}\phi^2. \tag{2.24}$$

The Hessian can be explicitly read from the Lagrangian as follows:

$$\mathcal{L} = \frac{1}{2}(\ddot{\phi}, \dot{\psi})\begin{pmatrix} 1 & a \\ a & 1 \end{pmatrix}\begin{pmatrix} \ddot{\phi} \\ \dot{\psi} \end{pmatrix} + \frac{1}{2}\phi^2. \tag{2.25}$$

So the theory is non-degenerate, only if $a \neq \pm 1$. If we compute the equations of motion, we get

$$[\text{EoM } \phi]: \quad \ddddot{\phi} + a\dddot{\psi} + \phi = 0, \quad [\text{EoM } \psi]: \quad \ddot{\psi} + a\dddot{\phi} = 0, \tag{2.26}$$

which can be combined in

$$(1 - a^2)\ddddot{\phi} = -\phi, \quad \ddot{\psi} = -a\dddot{\phi}. \tag{2.27}$$

Observe that we cannot solve for $\ddddot{\phi}$ and $\ddot{\psi}$ if $a = \pm 1$. In the non-degenerate case ($a \neq \pm 1$), we will need 4 initial conditions for ϕ and 2 for ψ, making a total of 3 dofs, one of which were hidden in the higher-order derivatives of ϕ.

2.3   Hamiltonian Procedure for Higher-Order Lagrangians. The Ostrogradski Theorem

To treat the general theory $L[\phi, \dot{\phi}, \ddot{\phi}]$ (which we assumed to be non-degenerate) of the previous section in the Hamiltonian language, one can make use of the Ostrogradski procedure [53].[5] As we will see, this method leads to a crucial result about stability in field theory, the *Ostrogradski theorem*.

Let us start with the construction of the Hamiltonian. Firstly, one should extract the information about the extra dof (i.e., the high-order derivatives) by introducing another pair of canonical variables in the Hamiltonian formalism. The canonical positions and momenta are defined as follows:

$$q_1 := \phi, \qquad q_2 := \dot{\phi}, \qquad \pi_1 := \frac{\delta L}{\delta \dot{\phi}} - \frac{d}{dt}\frac{\delta L}{\delta \ddot{\phi}}, \qquad \pi_2 := \frac{\delta L}{\delta \ddot{\phi}}. \qquad (2.28)$$

The non-degeneracy condition implies that one can invert these definitions to write

$$(\dot{q}_2 =) \quad \ddot{\phi} = F(q_1, q_2, \pi_2), \qquad (2.29)$$

such that the following equation is fulfilled:

$$\pi_2 = \frac{\delta L}{\delta \ddot{\phi}}\Big|_{\phi=q_1, \dot{\phi}=q_2, \ddot{\phi}=F}. \qquad (2.30)$$

Observe how the second-order derivatives $\ddot{\phi}$ become $\dot{q}_2$, i.e., we have hidden the higher-order-derivative nature of the Lagrangian by introducing a new variable. Now we can construct the Hamiltonian in the usual way, by performing the Legendre transformation of the Lagrangian with respect to the time derivatives of the fields (in this case $\{q_1, q_2\}$):

$$\begin{aligned}
H[q_1, q_2, \pi_1, \pi_2] &= \int \left[\dot{q}_1\pi_1 + \dot{q}_2\pi_2\right]d^3x - L[q_1, q_2, \dot{q}_2] \\
&\overset{(2.29)}{=} \int \left[q_2\pi_1 + F(q_1, q_2, \pi_2)\pi_2\right]d^3x - L[q_1, q_2, F(q_1, q_2, \pi_2)] \\
&\overset{(2.18)}{=} \int \left[q_2\pi_1 + F(q_1, q_2, \pi_2)\pi_2 \right. \\
&\qquad \left. - \mathcal{L}\big(q_1, \partial_i q_1, q_2, \partial_i \partial_j q_1, \partial_i q_2, F(q_1, q_2, \pi_2)\big)\right]d^3x. \quad (2.31)
\end{aligned}$$

Here an important thing to notice is that the momentum π_1 enters linearly, which implies that the Hamiltonian is unbounded from below in the direction of the phase

---

[5] Later we will see the case of higher-order time derivatives. Regarding the generalization to multi-fields, one just should apply the procedure to each of the fields of the theory.

space given by π_1. This is a particular case of a more general result called Ostrogradski theorem. Here we reproduce it as was stated in [2] (see also [32, 56]):

Theorem 2.1 (Ostrogradski 1850) *Let a Lagrangian involve n-th-order finite time derivatives of variables. If $n \geq 2$ and the Lagrangian is non-degenerate with respect to the highest-order derivatives, the Hamiltonian of this system linearly depends on a canonical momentum.*

This unboundedness implies the presence of ghosts (*Ostrogradski ghosts*) in the theory. Therefore, whenever we are dealing with a higher-order theory, one has to ensure that the theory is degenerate, in order to bypass the devastating consequences of this theorem.

Let us see a particular example of the previous derivation:

Example: The Ostrogradski procedure

Consider the following Lagrangian for a scalar field

$$\mathcal{L} = \frac{1}{2}\partial_\mu\phi\partial^\mu\phi + \frac{\lambda}{2}(\Box\phi)^2 - V(\phi), \qquad \lambda \neq 0. \tag{2.32}$$

where $\Box := \eta^{\mu\nu}\partial_\mu\partial_\nu$. The corresponding Lagrangian functional is

$$L[\phi, \dot{\phi}, \ddot{\phi}] = \int d^3x\mathcal{L} = \int d^3x \left[\frac{1}{2}(\dot{\phi}^2 - |\nabla\phi|^2) + \frac{\lambda}{2}(\ddot{\phi} - \Delta\phi)^2 - V(\phi)\right], \tag{2.33}$$

where Δ is the spatial Laplacian. This Lagrangian is non-degenerate, as one can easily check:

$$\det\left[\text{Hess}(L)\right] = \frac{\partial^2 L}{\partial\ddot{\phi}^2} = \lambda \quad \neq 0. \tag{2.34}$$

We start the Ostrogradski procedure by introducing the set of canonical variables

$$q_1 := \phi, \qquad \pi_1 := \frac{\delta L}{\delta\dot{\phi}} - \frac{d}{dt}\frac{\delta L}{\delta\ddot{\phi}} = \dot{\phi} - \lambda\dddot{\phi} + \lambda\Delta\dot{\phi},$$

$$q_2 := \dot{\phi}, \qquad \pi_2 := \frac{\delta L}{\delta\ddot{\phi}} = \lambda(\ddot{\phi} - \Delta\phi). \tag{2.35}$$

From the expression of π_2 we can obtain the function F,

$$\ddot{\phi} = F(q_1, q_2, \pi_2) = \frac{1}{\lambda}\pi_2 + \Delta q_1. \tag{2.36}$$

In this variables the Lagrangian density has the form

$$\mathcal{L} = -\frac{1}{2}(|\nabla q_1|^2 - (q_2)^2) + \frac{1}{2\lambda}(\pi_2)^2 - V(q_1). \tag{2.37}$$

The resulting Hamiltonian functional,

$$
\begin{aligned}
H[q_1, q_2, \pi_1, \pi_2] \\
= \int d^3x \left[\dot{q}_1 \pi_1 + \dot{q}_2 \pi_2 - \mathcal{L} \right] \\
= \int d^3x \left[q_2 \pi_1 + \left(\frac{1}{2\lambda} \pi_2 + \Delta q_1 \right) \pi_2 + \frac{1}{2} \left(|\nabla q_1|^2 - (q_2)^2 \right) + V(q_1) \right],
\end{aligned}
\tag{2.38}
$$

is indeed linear in the momentum the direction of π_1. Actually, one can separate the coupling between π's and q's, via the coordinate transformation ($\pi_1 \rightarrow \pi_1 + q_2/2$, $\pi_2 \rightarrow \pi_2 - \lambda \Delta q_1$) in phase space, which leads us to:

$$
H[q_1, q_2, \pi_1, \pi_2] = \int d^3x \left[q_2 \pi_1 + \frac{1}{2\lambda} (\pi_2)^2 + \frac{1}{2} \left(|\nabla q_1|^2 - \lambda (\Delta q_1)^2 \right) + V(q_1) \right].
\tag{2.39}
$$

Some Remarks

We will now highlight a couple of aspects about the Ostrogradski Hamiltonian:

1. The Hamilton equations of the Ostrogradski Hamiltonian reproduce the same dynamics as the original Lagrangian. Indeed, for the case we have been treating in this section, if we compute the four variations of the Hamiltonian,

$$
\frac{\delta H}{\delta \pi_1} = q_2,
\tag{2.40}
$$

$$
\frac{\delta H}{\delta \pi_2} = F + \pi_2 \frac{\partial F}{\partial \pi_2} - \frac{\delta L}{\delta \ddot{\phi}} \frac{\partial F}{\partial \pi_2} = F,
\tag{2.41}
$$

$$
\frac{\delta H}{\delta q_1} = \frac{\partial F}{\partial q_1} \pi_2 - \frac{\delta L}{\delta \phi} - \frac{\delta L}{\delta \ddot{\phi}} \frac{\partial F}{\partial q_1} = -\frac{\delta L}{\delta \phi},
\tag{2.42}
$$

$$
\frac{\delta H}{\delta q_2} = \pi_1 + \frac{\partial F}{\partial q_2} \pi_2 - \frac{\delta L}{\delta \dot{\phi}} - \frac{\delta L}{\delta \ddot{\phi}} \frac{\partial F}{\partial q_2} = \pi_1 - \frac{\delta L}{\delta \dot{\phi}},
\tag{2.43}
$$

one easily realizes that three of the Hamilton equations are nothing but the definitions of q_2, F and π_1, respectively,

$$
\dot{q}_1 = \frac{\delta H}{\delta \pi_1} \qquad \Leftrightarrow \qquad \dot{q}_1 = q_2,
\tag{2.44}
$$

$$
\dot{q}_2 = \frac{\delta H}{\delta \pi_2} \qquad \Leftrightarrow \qquad \dot{q}_2 = F(q_1, q_2, \pi_2),
\tag{2.45}
$$

$$
\dot{\pi}_2 = -\frac{\delta H}{\delta q_2} \qquad \Leftrightarrow \qquad \dot{\pi}_2 = -\pi_1 + \frac{\delta L}{\delta \dot{\phi}},
\tag{2.46}
$$

whereas the remaining one is the Euler-Lagrange equation (2.21):

$$\dot{\pi}_1 = -\frac{\delta H}{\delta q_1} \qquad \Leftrightarrow \qquad 0 = \frac{\delta L}{\delta \phi} - \dot{\pi}_1 = \frac{\delta L}{\delta \phi} - \frac{d}{dt}\left(\frac{\delta L}{\delta \dot{\phi}} - \frac{d}{dt}\frac{\delta L}{\partial \ddot{\phi}}\right). \quad (2.47)$$

2. If the theory is invariant under time translations, the conserved charge (the *energy* of the system) coincides with the Ostrogradski Hamiltonian [63].

Generalization to arbitrarily high derivatives

This procedure can be generalized to higher-order derivatives (see e.g. [63]). For a non-degenerate Lagrangian depending up to n-th derivatives of some field ϕ, one needs to introduce n new fields and momenta $\{q_i, \pi_i\}_{i=1}^n$ defined as

$$q_i := \phi^{(i-1)}, \qquad \pi_i := \sum_{j=i}^{n} \left(-\frac{d}{dt}\right)^{j-i} \frac{\delta L}{\delta \phi^{(j)}}. \quad (2.48)$$

For instance

$$q_1 := \phi, \qquad q_2 := \dot{\phi}, \qquad q_3 := \ddot{\phi}, \qquad \dots \qquad q_n := \phi^{(n-1)}, \quad (2.49)$$

$$\pi_n := \frac{\delta L}{\delta \phi^{(n)}}, \qquad \pi_{n-1} := \frac{\delta L}{\delta \phi^{(n-1)}} - \frac{d}{dt}\frac{\delta L}{\delta \phi^{(n)}},$$

$$\pi_{n-2} := \frac{\delta L}{\delta \phi^{(n-2)}} - \frac{d}{dt}\frac{\delta L}{\delta \phi^{(n-1)}} + \frac{d^2}{dt^2}\frac{\delta L}{\delta \phi^{(n)}}, \quad \dots \quad (2.50)$$

Then it is possible to solve from the n-th momentum (thanks to the degeneracy condition):

$$\phi^{(n)} = F(q_1, \dots, q_n, \pi_n) \quad \text{such that} \quad \pi_n = \frac{\delta L}{\delta \phi^{(n)}}\bigg|_{\phi=q_1,\dots,\phi^{(n-1)}=q_n,\phi^{(n)}=F}, \quad (2.51)$$

and the resulting Hamiltonian,

$$H[q_1, \dots, q_n, \pi_1, \dots \pi_n] = \int \left[q_2\pi_1 + q_3\pi_2 + \dots + q_n\pi_{n-1} + F(q_1, \dots, q_n, \pi_n)\pi_n\right]d^3x$$

$$- L[q_1, \dots, q_n, F(q_1, \dots, q_n, \pi_n)]. \quad (2.52)$$

is linear in $\{\pi_2, \dots, \pi_n\}$ (all momenta except π_1). These correspond to extra dofs of ghostly nature.

One can check that the Hamilton equation for π_1 is the one containing the Euler-Lagrange equation of the theory. The equations of the other momenta π_i give the definition of π_{i-1}, the equation of q_n gives the definition of F, and the equations of the remaining positions q_i will give simply $\dot{q}_i = q_{i+1}$ [63]. See that this clearly generalizes what we obtained in (2.44)–(2.46).

2.4 Explicit Kinetic Terms for the Ghosts via Auxiliary Fields

In some cases, it is possible to transform an "Ostrogradski" ghost (non-degenerate contributions to the Lagrangian with higher-order derivatives) into an "ordinary ghost" (some field with negative kinetic term with no higher-order derivatives). The technique is based on a very important concept in field theory (with many applications!): auxiliary fields.

Definition 2.3 (*Auxiliary field*) We call *auxiliary field* a field whose equation of motion is algebraic and admit a unique solution.

When a field is auxiliary, the solution of its equation of motion can be substituted back in the action without altering the dynamics. The Lagrangians containing auxiliary fields are quadratic polynomials in them. Therefore, the equations of motion that correspond to the auxiliary fields are first-degree equations with a unique solution.[6]

Let us revisit the previous example from a different perspective:

Example: An equivalent Lagrangian

First, we write down again the Lagrangian we used in the previous example,

$$\mathcal{L}_\phi = \frac{1}{2}\partial_\mu\phi\partial^\mu\phi + \frac{\lambda}{2}(\Box\phi)^2 - V(\phi), \qquad \lambda \neq 0. \tag{2.53}$$

Consider now the following Lagrangian density depending on two fields ϕ and χ

$$\mathcal{L}_{\phi\chi} = \frac{1}{2}\partial_\mu\phi\partial^\mu\phi + \chi\Box\phi - \frac{1}{2\lambda}\chi^2 - V(\phi). \tag{2.54}$$

We are going to show that the latter is equivalent to the former one.

One way of seeing this is by computing the set of equations of motion for both theories and check that they agree. But there is another way, which is based on the fact that the field χ is auxiliary. Its equation of motion can be solved as

$$0 = \frac{\partial \mathcal{L}_{\phi\chi}}{\partial \chi} = \Box\phi - \frac{1}{\lambda}\chi \quad \Rightarrow \quad \chi = \lambda\Box\phi, \tag{2.55}$$

---

[6] This definition is often relaxed to include any field whose equations of motion do not involve derivatives, which usually present several branches of solutions.

and this can be plugged back in the Lagrangian. This leads exactly to (2.53), finishing the proof.

In this example we have seen that in the reformulation of the theory in terms of $\{\phi, \chi\}$ the second-order derivatives do not appear. So, where is the ghost? The healthy dof and the extra ghost have been encoded in the pair $\{\phi, \chi\}$. Indeed, in order to see the ghost explicitly, one just need to perform an appropriate field redefinition (see also Appendix D of [42]):

Example: Kinetic term for the Ostrogradski ghost

Consider again

$$\mathcal{L}_{\phi\chi} = \frac{1}{2}\partial_\mu\phi\partial^\mu\phi + \chi\Box\phi - \frac{1}{2\lambda}\chi^2 - V(\phi). \tag{2.56}$$

After extracting a boundary term we get the following kinetic matrix:

$$\mathcal{L}_{\phi\chi} = \frac{1}{2}(\partial_\mu\phi, \partial_\mu\chi)\begin{pmatrix} 1 & -1 \\ -1 & 0 \end{pmatrix}\begin{pmatrix} \partial^\mu\phi \\ \partial^\mu\chi \end{pmatrix} - \frac{1}{2\lambda}\chi^2 - V(\phi) + \partial_\mu(\chi\partial^\mu\phi). \tag{2.57}$$

The determinant of the kinetic matrix is negative, so there is a ghost in the theory. Indeed, we can diagonalize it via a field redefinition ($\phi \mapsto \beta - \alpha$, $\chi \mapsto \beta$) and the Lagrangian becomes

$$\mathcal{L}_{\alpha\beta} = \frac{1}{2}\partial_\mu\alpha\partial^\mu\alpha - \frac{1}{2}\partial_\mu\beta\partial^\mu\beta - \frac{1}{2\lambda}\beta^2 - V(\beta - \alpha). \tag{2.58}$$

We can clearly see that the kinetic term of β has the wrong sign. This field is responsible for the Hamiltonian not to be bounded from below (it can identified with the Ostrogradski ghost of the original theory).

2.5 The Massive Spin-1

In order to describe the 3 dofs of a massive spin-1, the simplest Lorentz tensor that one can consider is a vector field A_μ. The most general kinetic term for it (with Lorentz invariance) is a linear combination of the objects

$$I_1 := \partial_\mu A_\nu \partial^\mu A^\nu, \qquad I_2 := \partial_\mu A_\nu \partial^\nu A^\mu, \qquad I_3 := (\partial_\mu A^\mu)^2. \tag{2.59}$$

I_2 and I_3 are linearly dependent up to boundary terms, so we can drop one of them (the I_2, for convenience). Moreover, we can consider the combination

$$F_{\mu\nu}(A)F^{\mu\nu}(A) = 2(\partial_\mu A_\nu \partial^\mu A^\nu - \partial_\mu A_\nu \partial^\nu A^\mu) = 2(I_1 - I_2) \qquad (2.60)$$

instead of I_1. Here we have introduced the antisymmetric combination $F_{\mu\nu}(A) := \partial_\mu A_\nu - \partial_\nu A_\mu$. The Lagrangian is then

$$\mathcal{L}_A = -\frac{1}{4}a F_{\mu\nu}(A)F^{\mu\nu}(A) - \frac{1}{2}b(\partial_\mu A^\mu)^2 + \frac{1}{2}m^2 A_\mu A^\mu, \qquad (2.61)$$

where we have introduced a mass term for the field ($m^2 > 0$), and two real parameters $a, b \in \mathbb{R}$.

This theory propagates a ghostly dof due to second term, and such a ghost can be transformed into an Ostrogradski one. To do that we perform a splitting of the vector into longitudinal (ϕ) and transversal (B_μ) as follows[7]

$$A_\mu = \partial_\mu \phi + B_\mu \quad \text{with} \quad \partial_\mu B^\mu = 0, \qquad (2.62)$$

which implies $\partial_\mu A^\mu = \Box\phi$ and $F_{\mu\nu}(A) = F_{\mu\nu}(B)$. After this redefinition, the Lagrangian density of the theory has the form

$$\mathcal{L}_{B\phi} = -\frac{1}{4}a F_{\mu\nu}(B)F^{\mu\nu}(B) - \frac{1}{2}b(\Box\phi)^2 + \frac{1}{2}m^2(\partial_\mu\phi + B_\mu)(\partial^\mu\phi + B^\mu) \quad (2.63)$$

$$= \frac{1}{2}a\delta^{ij}\left(\dot{B}_i\dot{B}_j + \partial_i B_0 \partial_j B_0 - 2\dot{B}_i\partial_j B_0\right) - a\delta^{k[i}\delta^{j]l}\partial_i B_j \partial_k B_l$$

$$- \frac{1}{2}b\left[\ddot{\phi}^2 + (\Delta\phi)^2 - 2\ddot{\phi}\Delta\phi\right]$$

$$+ \frac{1}{2}m^2(\dot{\phi}^2 - \delta^{ij}\partial_i\phi\partial_j\phi + (B_0)^2 - \delta^{ij}B_iB_j) + \underbrace{\partial_\mu(m^2 B^\mu\phi)}_{\text{boundary term}}. \qquad (2.64)$$

We require a to be non-trivial, because if not then we lose the dynamics for the vector part and only a scalar remains. Notice that B_0 does not propagate and that for $a \neq 0$ and $b \neq 0$ the Lagrangian functional is of the form $L_{B\phi}[B_0, B_i, \dot{B}_i, \phi, \dot{\phi}, \ddot{\phi}]$. It is actually enough to check the following determinant to realize that the theory has an Ostrogradski ghost:

$$\det\begin{pmatrix} \frac{\partial^2 \mathcal{L}_{B\phi}}{\partial \dot{B}_i \partial \dot{B}_j} & \frac{\partial^2 \mathcal{L}_{B\phi}}{\partial \dot{B}_i \partial \ddot{\phi}} \\ \frac{\partial^2 \mathcal{L}_{B\phi}}{\partial \ddot{\phi}\partial \dot{B}_j} & \frac{\partial^2 \mathcal{L}_{B\phi}}{\partial \ddot{\phi}\partial \ddot{\phi}} \end{pmatrix} = \det\begin{pmatrix} a\mathbb{1}_3 & 0_{1\times 3} \\ 0_{3\times 1} & -b \end{pmatrix} = -a^3 b. \qquad (2.65)$$

The only way to avoid the Ostrogradski ghost (and maintain the vector dofs of B_i) is by requiring $b = 0$. Notice that a should also be positive in order to avoid a wrong sign in the kinetic term of B_i. So, from now on, we take $b = 0$ and $a > 0$. Moreover,

---

[7] This decomposition can always be done. See [60, Sect. 8.7.1].

we can redefine the fields and the mass parameter absorbing this a in order to get the canonical normalization for the kinetic term of the vector part, so that (2.61) becomes:

$$\mathcal{L}_{\text{Proca}} = -\frac{1}{4} F_{\mu\nu}(A) F^{\mu\nu}(A) + \frac{1}{2} m^2 A_\mu A^\mu . \qquad (2.66)$$

This is the *Proca Lagrangian* and propagates the three dofs needed for the quantum field theory of a massive spin-1 particle: two in the transversal part of the field B_μ and one in the longitudinal part ϕ. Observe that the dynamics of ϕ is controlled by the mass term. In the massless case (Maxwell) such a dof is absent.

It is also interesting to notice that the Proca Lagrangian is completely equivalent to the so called *Stückelberg Lagrangian* (after the introduction of a new field $A_\mu \to A_\mu + \partial_\mu \varphi$):

$$\mathcal{L}_{\text{Stück}} = -\frac{1}{4} F_{\mu\nu}(A) F^{\mu\nu}(A) + \frac{1}{2} m^2 (\partial_\mu \varphi + A_\mu)(\partial^\mu \varphi + A^\mu) . \qquad (2.67)$$

This Lagrangian has a local symmetry

$$A_\mu \to A_\mu + \partial_\mu g , \qquad \varphi \to \varphi - g , \qquad (2.68)$$

so any given φ is redundant since it can be totally absorbed by choosing an appropriate g. If we fix the gauge in which $\varphi = 0$, we recover the Proca Lagrangian, so $\mathcal{L}_{\text{Stück}}$ also describes the 3 dofs of a massive spin-1. In the Stücekelberg case, the gauge symmetry is telling us that the longitudinal mode that we can extract from A_μ does not introduce ghosts, because it can be absorbed in φ, which has a healthy kinetic term. The Proca field then has 1 longitudinal dof (thanks to the non-vanishing mass) and 2 transversal dofs.

It is interesting to notice how the gauge symmetries are connected with the absence of Ostrogradski ghosts coming from the longitudinal part of the fields. In the massless case, the U(1) gauge symmetry $A_\mu \to A_\mu + \partial_\mu g$ prevents the Maxwell Lagrangian from being pathological. The same thing happens in the Fierz-Pauli Lagrangian for a massless spin-2 field, where the ghosts are absent and the kinetic sector corresponds to the one with the gauge symmetry $h_{\mu\nu} \to h_{\mu\nu} + \partial_\mu \xi_\nu + \partial_\nu \xi_\mu$. Observe that, in these two cases, what the gauge freedom is telling us is that the longitudinal parts (the ones connected to the Ostrogradski instability) are pure gauge, i.e. redundant/unphysical. In other words, they do not appear in the Lagrangian, proving that the theories are safe from these problems.

Chapter 3
Examples of Instabilities in Gravity Theories

Abstract In the third chapter, we will briefly present general results on the appearance of instabilities in modified gravity. We will use the techniques developed in the previous chapters to unveil the presence of instabilities and discuss how to avoid them.

3.1 Stability Issues in Modified Theories of Gravity

In this lecture we shall use the knowledge of the previous chapters to analyze the possible appearance of instabilities in modifications of General Relativity (GR). A Hamiltonian analysis of the Einstein-Hilbert action reveals that the theory is healthy at the full nonlinear level [55]. Nevertheless, when modifying such an action, we may expect some pathological behavior. Let us explain the reasoning behind this in the following.

The Lovelock theorem states that: "The only possible equations of motion from a local gravitational action which contains only second derivatives of the four-dimensional spacetime metric are the Einstein field equations". From such a statement we can derive a straightforward conclusion: if we make a local, four-dimensional modification of GR, we are introducing higher derivatives and/or extra fields (scalar, vector and tensors). On the one hand, from the previous lecture we know that having higher-than-second-order time derivatives in the action almost certainly signals a pathological behavior. On the other hand, the extra fields present in the action could be ghosts or tachyons depending on the parameters of the theory.

Therefore, when exploring possible modifications of GR, instabilities of different kinds are expected. Here we present some examples of modified theories of gravity and their known pathological/safe behavior. We follow the discussion in Sect. 7.2 of [40]:

- First, we discuss about metric theories with higher-order terms in the curvature. It is known that only Lovelock theories have generically the same spectrum as General Relativity, i.e., a massless graviton. Any other higher-order curvature

© The Author(s), under exclusive license to Springer Nature Switzerland AG 2023
A. Delhom et al., *Instabilities in Field Theory*, SpringerBriefs
in Physics, https://doi.org/10.1007/978-3-031-40433-7_3

theory propagates additional typically problematic dofs (generically, a massive ghostly spin-2 and a scalar). An example of this is (Cosmological) Einsteinian Cubic gravity, as we will see in Sect. 3.3, which contains a ghost that becomes strongly coupled around spatially-flat cosmological backgrounds [11].

However, a miracle happens in some degenerate cases, and these generic problems are absent. A good example of this is the $f(\mathring{R})$ family of theories. This can be seen directly in the Einstein frame (the result is just GR plus a scalar field) or by taking a Lagrangian of the type $\mathring{R}^2 + a\mathring{R}_{\mu\nu}\mathring{R}^{\mu\nu}$ and analyzing the limit $a \to 0$. It can be seen that the mass of the extra spin-2 goes to infinity without rendering any pathologies [11]. This result for $\mathring{R}^2$ can be extended to the entire family of $f(\mathring{R})$ [64].

Another miraculous case is a Lagrangian consisting on an arbitrary function of the Gauss-Bonnet invariant $\mathcal{G}$, called $f(\mathcal{G})$ gravity (see e.g. [22, 25, 51]). Here again, only the massless graviton and a safe scalar propagate [44].

- Ricci-based gravity theories with projective symmetry (see Sect. 3.4) are known to be equivalent GR up to a field redefinition (i.e., they admit an Einstein frame), so only the graviton propagates. However, when the action also includes the anti-symmetric part of the Ricci tensor, the broken projective symmetry leads to new ghostly dofs [6, 7]. We will discuss this in more detail in Sect. 3.4.

- The nonlinear extensions of the teleparallel equivalents, $f(T)$ [14, 34, 45] and $f(Q)$ [8, 9] also suffer from instabilities. The key to understand the origin of these new dofs is the following: the teleparallel equivalents of GR, T and Q exhibit some special symmetries that lead to a propagating graviton. To be precise, such groups of transformations only leave the action invariant up to a boundary term. Consequently, these boundary terms cannot escape from the function f in the nonlinear extension. The symmetries are then lost, and new dofs appear.

  For instance, it has be shown that maximally symmetric backgrounds are strongly coupled in $f(Q)$ [9]. For $f(T)$, the situation is even worse, since the strong coupling problem affects more general cosmological backgrounds [34].

- The quadratic Poincaré Gauge (PG) gravity Lagrangian, which will be explored in Sect. 3.2, contains ghosts and tachyons for generic values of the parameters (see e.g. [15]). In fact, in [65, 66] (and more recently in [12]), authors found that the only modes that could propagate safely were the two spin-0 with different parity. Similar problems are expected in quadratic metric-affine gravity (see for example the recent work [41]).

All of the examples above are summarized in Table 3.1 for the healthy theories, and in Table 3.2 for the unstable ones.

In the following sections we shall study in detail the appearance of instabilities in PG gravity [12], (Cosmological) Einsteinian Cubic gravity [11], and Ricci-based theories [6, 7].

Table 3.1 Some examples of modified theories of gravity which do not introduce instabilities under some restrictions

Theory	Field content
Lovelock (GR, GB,...)	Graviton
$f(\mathring{R})$, $f(\text{GB})$	Graviton + scalar
$f(R)$	Graviton [+ non-dynamical scalar]
Horndeski gravity	Graviton + scalar
Generalized Proca	Graviton + (massive) vector
Ricci-Based Gravity	Graviton + scalar

Table 3.2 Paradigmatic examples of modifications of GR and their known pathological behavior

Theory	Some known pathologies
Massive gravity (original formulation)	Boulware-Deser ghost
Ricci-Based Gravity with $R_{[\mu\nu]}$	Ghost (projective mode)
Generic higher curvature gravity	Potential ghosts (massive spin-2, scalar)
$f(T)$	Strong coupling problem in FLRW
$f(Q)$	Strong coupling problem in Max. Sym.
General quadratic PG gravity	Ghosts and tachyons (and strong c.)
General quadratic MAG gravity	Similar problems as in PG

3.2 Stable Restrictions of Quadratic Poincaré Gauge Gravity

With the knowledge acquired in the previous lectures we are ready to study the stability of PG gravity. We shall summarize the results of [12], where the stable modes of propagation of the action quadratic PG action were found. We shall unveil the presence of pathological terms in a background-independent approach just by looking at the interactions of the different torsion components.

To start with, let us present the action of quadratic PG gravity, which is given by

$$
\begin{aligned}
S_{\text{PG}} = \int \mathrm{d}^4 x \sqrt{|g|} \, \big(& a_0 R + a_1 T_{\mu\nu\rho} T^{\mu\nu\rho} + a_2 T_{\mu\nu\rho} T^{\nu\rho\mu} + a_3 T^{\rho\mu}{}_\rho T^\nu{}_{\mu\nu} + b_1 R^2 \\
& + b_2 R_{\mu\nu\rho\sigma} R^{\mu\nu\rho\sigma} + b_3 R_{\mu\nu\rho\sigma} R^{\rho\sigma\mu\nu} + b_4 R_{\mu\nu\rho\sigma} R^{\mu\rho\nu\sigma} \\
& + b_5 R_{\mu\nu} R^{\mu\nu} + b_6 R_{\mu\nu} R^{\nu\mu} \big) ,
\end{aligned}
\tag{3.1}
$$

where the a_i and b_i are constant parameters of the theory, the torsion is given by $T^\rho{}_{\mu\nu} := 2\Gamma^\rho{}_{[\mu\nu]}$ and the curvatures are those of the torsionful connection $\Gamma^\rho{}_{\mu\nu}$. This theory has clearly more dofs than GR, due to the quadratic curvature and torsion terms. To analyse those introduced by the torsion tensor, it is customary to decompose the torsion into three irreducible pieces:

$$\begin{cases} \text{Trace vector: } T_\mu := T^\nu{}_{\mu\nu}, \\[2mm] \text{Axial vector: } S_\mu := \varepsilon_{\mu\nu\rho\sigma} T^{\nu\rho\sigma}, \\[2mm] \text{Tensor } \mathrm{t}^\rho{}_{\mu\nu}, \text{ such that } \mathrm{t}^\nu{}_{\mu\nu} = 0 \text{ and } \varepsilon_{\mu\nu\rho\sigma}\mathrm{t}^{\nu\rho\sigma} = 0, \end{cases} \quad (3.2)$$

such that

$$T^\rho{}_{\mu\nu} = \frac{1}{3}\left(T_\mu \delta^\rho_\nu - T_\nu \delta^\rho_\mu\right) + \frac{1}{6}\varepsilon^\rho{}_{\mu\nu\sigma} S^\sigma + \mathrm{t}^\rho{}_{\mu\nu}. \tag{3.3}$$

This decomposition turns out to be very useful, thanks to the fact that the three terms in (3.2) propagate different dynamical off-shell dofs.

Let us also remember that every non-symmetric and metric compatible connection, it can be related with the Levi-Civita connection as[1]

$$\Gamma^\alpha{}_{\mu\nu} = \mathring{\Gamma}^\alpha{}_{\mu\nu} + K^\alpha{}_{\mu\nu}, \tag{3.4}$$

where $\mathring{\Gamma}$ is the Levi-Civita connection and

$$K^\alpha{}_{\mu\nu} := \frac{1}{2}\left(T^\alpha{}_{\mu\nu} + T_\mu{}^\alpha{}_\nu + T_\nu{}^\alpha{}_\mu\right) \tag{3.5}$$

is the *contorsion tensor*. The decomposition of the quantities related to the total connection using (3.4) is known as a *post-Riemannian expansion*.

With respect to the stability of the Lagrangian (3.1), in order to avoid ghosts already for the graviton when the torsion is set to zero, we will impose the recovery of the Gauss-Bonnet term in the limit of vanishing torsion. In 4 dimensions we can use the topological nature of the Gauss-Bonnet term to remove one of the parameters. More explicitly, we have

$$\mathcal{L}_{\mathrm{PG}}\big|_{T=0} = a_0\mathring{R} + \left(b_2 + b_3 + \frac{b_4}{2}\right)\mathring{R}_{\mu\nu\rho\sigma}\mathring{R}^{\mu\nu\rho\sigma} + (b_5 + b_6)\mathring{R}_{\mu\nu}\mathring{R}^{\mu\nu} + b_1\mathring{R}^2, \tag{3.6}$$

so the Gauss-Bonnet term for the quadratic sector is recovered upon requiring

$$b_5 = -4b_1 - b_6, \quad b_4 = 2(b_1 - b_2 - b_3). \tag{3.7}$$

In the following subsection we shall show that imposing stability in the torsion vector modes reduces drastically the parameter space of PG gravity.

---

[1] Note that, though the connection symbols are not tensors, the difference of two connections is always a tensor field.

3.2.1 Ghosts in the Vector Sector

In 4 dimensions, a vector field A_μ has four components: one temporal A_0, and three spatial A_i, with $i = 1, 2, 3$. However, as we saw in the previous lecture, they cannot propagate at the same time without introducing a ghostly dof. In particular for any theory describing a massive vector, like the ones present in the PG action, we must require the following conditions in order to avoid ghosts [35, 36]:

- **The equations of motion must be of second order**. As we explained in the previous lecture, this is because the Ostrogradski theorem predicts ghosts for higher-order equations of motion.
- **The temporal component of the vector field A_0 should not be dynamical**. Therefore, the massive vector under this ghost-free condition would only propagate three dofs, which is exactly the ones that the massive spin-1 representation of the Lorentz group can propagate.

Following these prescriptions, in this subsection we shall constrain the parameter space of PG gravity by imposing stability in the two massive spin-1 fields that are part of the particle spectrum of this theory.

In order to do so, we look at the vector sector containing the trace T_μ and the axial component S_μ of the torsion, while ignoring the pure tensor part $\mathsf{t}^\rho{}_{\mu\nu}$ for the moment. Plugging the decompositions (3.4) and (3.3) into the quadratic PG Lagrangian (3.1) we obtain

$$
\begin{aligned}
\mathcal{L}_\mathrm{v} = {} & -\frac{2}{9}(\kappa - \beta)\mathcal{T}_{\mu\nu}\mathcal{T}^{\mu\nu} + \frac{1}{72}(\kappa - 2\beta)\mathcal{S}_{\mu\nu}\mathcal{S}^{\mu\nu} \\
& + \frac{1}{2}m_T^2 T_\mu T^\mu + \frac{1}{2}m_S^2 S_\mu S^\mu + \frac{\beta}{81} S_\mu S^\mu T_\nu T^\nu \\
& + \frac{4\beta - 9b_2}{81}\left[(S_\mu T^\mu)^2 + 3 S^\mu S^\nu \mathring{\nabla}_\mu T_\nu\right] + \frac{\beta}{54} S_\nu S^\nu \mathring{\nabla}_\mu T^\mu + \frac{\beta - 3b_2}{9} S^\mu T^\nu \mathring{\nabla}_\mu S_\nu \\
& + \frac{\beta - 3b_2}{12}(\mathring{\nabla}_\mu S^\mu)^2 + \frac{\beta}{36}\left(2\mathring{G}^{\mu\nu} S_\mu S_\nu + \mathring{R} S_\mu S^\mu\right),
\end{aligned}
\tag{3.8}
$$

where $\mathring{G}^{\mu\nu}$ is the Levi-Civita Einstein tensor, $\mathcal{T}_{\mu\nu} := 2\partial_{[\mu}T_{\nu]}$ and $\mathcal{S}_{\mu\nu} := 2\partial_{[\mu}S_{\nu]}$ are the field strengths of the trace and axial vectors respectively and we have defined

$$
\kappa := 4b_1 + b_6, \tag{3.9}
$$

$$
\beta := b_1 + b_2 - b_3, \tag{3.10}
$$

$$
m_T^2 := -\frac{2}{3}(2a_0 - 2a_1 + a_2 - 3a_3), \tag{3.11}
$$

$$
m_S^2 := \frac{1}{12}(a_0 - 4a_1 - 4a_2). \tag{3.12}
$$

In order to arrive at the final expression (3.8) we have used the Bianchi identities to eliminate terms containing $\mathring{R}_{\mu\nu\rho\sigma}\varepsilon^{\mu\nu\rho\sigma}$ and express $\mathring{R}_{\mu\nu\rho\sigma}\mathring{R}^{\mu\rho\nu\sigma} = \frac{1}{2}\mathring{R}_{\mu\nu\rho\sigma}\mathring{R}^{\mu\nu\rho\sigma}$.

We have also dropped the Gauss-Bonnet invariant of the Levi-Civita connection and the total derivative $\varepsilon_{\mu\nu\alpha\beta}S^{\mu\nu}\mathcal{T}^{\alpha\beta}$. Moreover, we have made a few integrations by parts and used the commutator of covariant derivatives. We can see that the parameter b_1 does not play any role since it simply corresponds to the irrelevant Gauss-Bonnet coupling constant.

If we look at the pure trace sector T_μ of (3.8), we observe that it does not contain non-minimal couplings or divergence-square terms. Unlike the torsion trace, the axial component S_μ shows very worrisome terms that appear in the three following ways:

- The first pathological term is $(\mathring{\nabla}_\mu S^\mu)^2$ that introduces a ghostly dof because it clearly makes the four components of the vector propagate. We shall get rid of it by imposing $\beta = 3b_2$.
- The non-minimal couplings to the curvature which are not Horndeski-like are known to lead to ghostly dofs. The presence of these instabilities shows in the metric field equations where again all the components of the vector will enter dynamically, hence revealing its problematic dynamics. An exception is the coupling to the Einstein tensor that avoids generating these second time derivatives due to its divergenceless property. For this reason we have explicitly separated the non-minimal coupling to the Einstein tensor in (3.8). It is therefore clear that we need to impose the additional constraint $\beta = 0$ to guarantee the absence ghosts, which, in combination with the above condition $\beta = 3b_2$, results in $\beta = b_2 = 0$.
- Furthermore, there are other interactions in (3.8) with a generically pathological character schematically given by $S^2 \nabla T$ and $ST \nabla S$. Although these may look like safe vector Galileon-like interactions, actually the fact that they contain both sectors makes them dangerous. This can be better understood by introducing Stückelberg fields, so we effectively have $T_\mu \to \partial_\mu \phi$ and $S_\mu \to \partial_\mu \psi$ with ϕ and ψ being the scalar and pseudo-scalar Stückelbergs. The interactions become of the form $(\partial\psi)^2\partial^2\phi$ and $\partial\phi\partial\psi\partial^2\phi$ that, unlike the pure Galileon interactions, generically give rise to higher-order equations of motion and, therefore, Ostrogradski instabilities. Nevertheless, we can see that the avoidance of this pathological behavior does not introduce new constraints on the parameters, since the coefficients in front of them in (3.8) are already zero if we take into account the two previous stability considerations.

Let us note that the extra constraint $\beta = 0$ genuinely originates from the quadratic curvature interactions in the PGT Lagrangian. Such interactions induce the non-minimal couplings between the axial sector and the graviton, as well as the problematic non-gauge-invariant derivative interactions. Observe that this constraint cannot be obtained from a perturbative analysis on a Minkowski background because, in that case, these interactions will only enter at cubic and higher-orders so that the linear analysis is completely oblivious to it.

We can see that the stability conditions not only remove the obvious pathological interactions mentioned before, but they actually eliminate all the interactions and only leave the free quadratic part

$$\mathcal{L}_v\big|_{b_2,\beta=0} = -\frac{2}{9}\kappa\mathcal{T}_{\mu\nu}\mathcal{T}^{\mu\nu} + \frac{1}{2}m_T^2 T_\mu T^\mu + \frac{1}{72}\kappa\mathcal{S}_{\mu\nu}\mathcal{S}^{\mu\nu} + \frac{1}{2}m_S^2 S_\mu S^\mu\,, \quad (3.13)$$

where we see that the kinetic terms for T_μ and S_μ have the same normalization but with opposite signs, hence leading to the unavoidable presence of a ghost. Therefore, the only stable possibility is to exactly cancel both kinetic terms. Consequently, the entire vector sector becomes non-dynamical.

Now that we have shown that the vector sector must trivialize in stable PGTs, we can return to the full torsion scenario by including the pure tensor sector $\mathsf{t}^\rho{}_{\mu\nu}$. Instead of using the general decomposition (3.3), it is more convenient to work with the torsion directly for our purpose here. We can perform the post-Riemannian decomposition for the theories with a stable vector sector to obtain

$$\mathcal{L}_{\text{stable}} = a_0\mathring{R} + b_1\mathcal{G} + a_1 T_{\mu\nu\rho}T^{\mu\nu\rho} + a_2 T_{\mu\nu\rho}T^{\nu\rho\mu} + a_3 T_\mu T^\mu\,. \quad (3.14)$$

The first term is just the usual Einstein-Hilbert Lagrangian, modulated by a_0, while the second term corresponds to the topological Gauss-Bonnet invariant for a connection with torsion, so we can safely drop it in four dimensions and, consequently, the first two terms in the above expression simply describe GR. The rest of the expression clearly shows the non-dynamical nature of the full torsion so that having a stable vector sector also eliminates the dynamics for the tensor component, therefore making the full connection an auxiliary field.

3.2.2   Constructing a Stable Poincaré Gauge Theory

The precedent subsection has been devoted to showing the presence of ghosts in general quadratic PG, when at least one of the spin-1 fields in T_μ or S_μ is propagating. Although this is a drawback for a very general class of theories, we can construct safe theories if we make these spin-1 fields non-propagating. In particular, in the following we will show a specific class of ghost-free theory which propagates the pseudo-scalar mode given by the longitudinal part (spin-0) of the axial torsion vector.

We can ask whether there is some non-trivial healthy theory described by (3.8) where the scalar is associated to the axial vector. The answer is indeed affirmative, and in order to prove such a result we simply need to impose the vanishing of the Maxwell kinetic terms that results in the following conditions

$$\kappa = 0 \quad \text{and} \quad \beta = 0\,. \quad (3.15)$$

Imposing these conditions, performing a few integrations by parts and dropping the Gauss-Bonnet term, the Lagrangian then reads

$$\mathcal{L}_{\text{Holst}} = a_0 \mathring{R} + \frac{1}{2} m_T^2 T_\mu T^\mu + \frac{1}{2} m_S^2 S_\mu S^\mu$$
$$+ \alpha \left[(\mathring{\nabla}_\mu S^\mu)^2 - \frac{4}{3} S_\mu T^\mu \mathring{\nabla}_\nu S^\nu + \frac{4}{9} (S_\mu T^\mu)^2 \right], \qquad (3.16)$$

with $\alpha := -\frac{b_2}{4}$.

At this moment, let us use the definition of the so-called Holst term,[2] which is given by $\mathcal{H} := \varepsilon^{\mu\nu\rho\sigma} R_{\mu\nu\rho\sigma}$ and whose post-Riemannian expansion is

$$\mathcal{H} = \frac{2}{3} S_\mu T^\mu - \mathring{\nabla}_\mu S^\mu, \qquad (3.17)$$

where we have used that $\varepsilon^{\mu\nu\rho\sigma} \mathring{R}_{\mu\nu\rho\sigma} = 0$ by virtue of the Bianchi identities. Thus, it is obvious that the Lagrangian can be expressed as

$$\mathcal{L}_{\text{Holst}} = a_0 \mathring{R} + \frac{1}{2} m_T^2 T_\mu T^\mu + \frac{1}{2} m_S^2 S_\mu S^\mu + \alpha \mathcal{H}^2. \qquad (3.18)$$

We will understand the nature of this scalar by introducing an auxiliary field ϕ to rewrite (3.18) as

$$\mathcal{L}_{\text{Holst}} = a_0 \mathring{R} + \frac{1}{2} m_T^2 T_\mu T^\mu + \frac{1}{2} m_S^2 S_\mu S^\mu - \alpha \phi^2 + 2\alpha \phi \varepsilon^{\mu\nu\rho\sigma} R_{\mu\nu\rho\sigma}. \qquad (3.19)$$

As we shall show now, the field ϕ is dynamical and corresponds to the pseudo-scalar mode in the PG Lagrangian.

At this moment, we can introduce the post-Riemannian expansion (3.17) into the Lagrangian, obtaining

$$\mathcal{L}_{\text{Holst}} = a_0 \mathring{R} + \frac{1}{2} m_T^2 T_\mu T^\mu + \frac{1}{2} m_S^2 S_\mu S^\mu - \alpha \phi^2 + 2\alpha \phi \left(\frac{2}{3} S_\mu T^\mu - \mathring{\nabla}_\mu S^\mu \right). \qquad (3.20)$$

The corresponding equations for S^μ and T^μ are

$$m_S^2 S_\mu + \frac{4\alpha\phi}{3} T_\mu + 2\alpha \partial_\mu \phi = 0, \qquad (3.21)$$

$$m_T^2 T_\mu + \frac{4\alpha\phi}{3} S_\mu = 0, \qquad (3.22)$$

respectively. For $m_T^2 \neq 0$,[3] we can algebraically solve these equations as

---

[2] Although this term is commonly known as the Holst term, due to the research article of Soren Holst in 1995 [38], in the context of torsion gravity it was first introduced by R. Hojman *et. al.* in 1980 [37].

[3] The singular value $m_T^2 = 0$ leads to uninteresting theories where all the dynamics is lost so we will not consider it any further here.

$$T_\mu = -\frac{4\alpha\phi}{3m_T^2}S_\mu, \tag{3.23}$$

$$S_\mu = -\frac{2\alpha\,\partial_\mu\phi}{m_S^2 - \left(\frac{4\alpha\phi}{3m_T}\right)^2}, \tag{3.24}$$

which we can plug into the Lagrangian to finally obtain

$$\mathcal{L}_{\text{Holst}} = a_0\mathring{R} - \frac{2\alpha^2}{m_S^2 - \left(\frac{4\alpha\phi}{3m_T}\right)^2}\partial_\mu\phi\partial^\mu\phi - \alpha\phi^2. \tag{3.25}$$

This equivalent formulation of the theory where all the auxiliary fields have been integrated out explicitly shows the presence of a propagating pseudo-scalar field. Moreover, we can see how including the pure tensor part $t^\rho{}_{\mu\nu}$ into the picture does not change the conclusions because it contributes to the Holst term as

$$\mathcal{H} = \frac{2}{3}S_\mu T^\mu - \mathring{\nabla}_\mu S^\mu + \frac{1}{2}\varepsilon_{\alpha\beta\mu\nu}t_\lambda{}^{\alpha\beta}t^{\lambda\mu\nu}. \tag{3.26}$$

This shows that $q_{\rho\mu\nu}$ only enters as an auxiliary field whose equation of motion trivializes it.

The stability constraints on the parameters can now be obtained very easily. From (3.25) we can realize that α must be positive to avoid having an unbounded potential from below. On the other hand, the condition to prevent ϕ from being a ghost depends on the signs of m_S^2 and m_T^2, which are not defined by any stability condition so far. We can distinguish the following possibilities:

- $m_S^2 > 0$: We then need to have $1 - \left(\frac{4\alpha\phi}{3m_Tm_S}\right)^2 > 0$. For $m_T^2 < 0$ this is always satisfied, while for $m_T^2 > 0$ there is an upper bound for the value of the field given by $|\phi| < |\frac{3m_Sm_T}{4\alpha}|$.

- $m_S^2 < 0$: The ghost-freedom condition is now $1 - \left(\frac{4\alpha\phi}{3m_Tm_S}\right)^2 < 0$, which can never be fulfilled if $m_T^2 > 0$. If $m_T^2 < 0$ we instead have the lower bound $|\phi| > |\frac{3m_Sm_T}{4\alpha}|$.

3.3 Strong Coupling in Cosmological Einsteinian Cubic Gravity

This section is devoted to a particular kind of cubic (metric) theory of gravity and the presence of instabilities due to strongly coupled dofs. To start with, we revise the definition of these theories, as well as some basic properties that are needed for our analysis.

3.3.1 Introduction to Einsteinian Cubic Graviti(es)

Einsteinian Cubic Gravity [16] is a higher-order curvature theory of gravity that possesses the same linear spectrum as General Relativity around maximally symmetric spacetimes in arbitrary dimension [16, 50, 52] (i.e., only the graviton propagates). These theories were extended in [4] to the so-called *Cosmological Einsteinian Cubic Gravity* (CECG), whose linear spectrum coincides also with the one of GR not only around maximally symmetric spaces but in arbitrary FLRW spacetimes (see e.g. [3, 4, 20]).

As it is well-known, Lovelock terms are the only higher-order curvature theories with the same field content as GR around arbitrary backgrounds [46, 47]. Consequently, the cubic theories we mentioned above must contain additional dofs, which will be connected to Ostrogradski instabilities due to the higher-order nature of the equations. Einsteinian Cubic gravities are specific cubic polynomial of the Riemann tensor such that the field equations become second order around specific backgrounds. Therefore, less dofs are propagating there, what is an indicator of a strong coupling problem. In this section, based on the work [11], we revise the implications of such a problem[4] around spatially flat cosmological models in CECG.

The Lagrangian we are going to analyze is

$$S[g_{\mu\nu}] = \int d^4x\sqrt{|g|}\left(-\Lambda + \frac{M_{\text{Pl}}{}^2}{2}\mathring{R} + \frac{\beta}{M_{\text{Pl}}{}^2}\mathcal{R}_{(3)}\right), \qquad (3.27)$$

where the General Relativity part has been corrected with the CECG invariant:[5]

$$\mathcal{R}_{(3)} := -\frac{1}{8}\left(12\mathring{R}_\mu{}^\rho{}_\nu{}^\sigma\mathring{R}_\rho{}^\tau{}_\sigma{}^\eta\mathring{R}_\tau{}^\mu{}_\eta{}^\nu + \mathring{R}_{\mu\nu}{}^{\rho\sigma}\mathring{R}_{\rho\sigma}{}^{\tau\eta}\mathring{R}_{\tau\eta}{}^{\mu\nu}\right.$$
$$+ 2\mathring{R}\mathring{R}_{\mu\nu\rho\sigma}\mathring{R}^{\mu\nu\rho\sigma} - 8\mathring{R}^{\mu\nu}\mathring{R}_\mu{}^{\rho\sigma\tau}\mathring{R}_{\nu\rho\sigma\tau} + 4\mathring{R}^{\mu\nu}\mathring{R}^{\rho\sigma}\mathring{R}_{\mu\rho\nu\sigma}$$
$$\left. - 4\mathring{R}\mathring{R}_{\mu\nu}\mathring{R}^{\mu\nu} + 8\mathring{R}_\mu{}^\nu\mathring{R}_\nu{}^\rho\mathring{R}_\rho{}^\mu\right). \qquad (3.28)$$

Let us insist one more time that this combination has the same linear spectrum as GR around cosmological backgrounds (i.e., only the graviton propagates).

For a FLRW universe of the type

$$ds^2 = dt^2 - a^2(t)\delta_{ij}dx^i dx^j, \qquad (3.29)$$

the gravitational Friedmann equation is

$$3M_{\text{Pl}}{}^2 H^2 - 6\frac{\beta}{M_{\text{Pl}}{}^2}H^6 = \rho + \Lambda, \qquad (3.30)$$

---

[4] These problems were already discussed for the case of inflationary solutions in [57].

[5] We keep the little ring over the curvatures to emphasize that we are in purely metric gravity, i.e., that all the curvatures are evaluated in the Levi-Civita connection.

where ρ is the energy-density of the matter sector and $H(t) := \frac{\dot{a}(t)}{a(t)}$. In the absence of matter, $\rho = 0$, we can see that we have (at most) three branches of expanding de Sitter solutions. For metric perturbations $g_{ij} = -a^2(\delta_{ij} + h_{ij})$ with h_{ij} transverse and traceless, we find the quadratic action [57]:

$$S^{(2)} = \frac{M_{\mathrm{Pl}}{}^4 - 6H_0^4\beta}{8M_{\mathrm{Pl}}{}^2} \sum_\lambda \int dt \, d^3x \, a^3 \left[\dot{h}_\lambda^2 - \frac{1}{a^2}(\partial_i h_\lambda)^2 \right], \qquad (3.31)$$

where the sum extends to the two polarizations of the gravitational waves and H_0 is the considered de Sitter branch. It is clear that $M_{\mathrm{Pl}}{}^4 - 6H_0^4\beta > 0$ is required to avoid ghostly gravitational waves and, consequently, the only allowed de Sitter solutions verify $\Lambda > 2H_0^2 > 0$ [57].

3.3.2 Generalities About Bianchi I Spaces

To prove that spatially flat FLRW solutions are singular in CECG, we make use of a general Bianchi I metric. These spaces are anisotropic extensions of the flat FLRW spaces (the *isotropic limit*):

$$ds^2 = \bar{g}_{\mu\nu}dx^\mu dx^\nu = N^2(t)dt^2 - a^2(t)dx^2 - b^2(t)dy^2 - c^2(t)dz^2, \qquad (3.32)$$

where $N(t)$ is the lapse function and $a(t)$, $b(t)$ and $c(t)$ stand for the scale factors along the three coordinate axis. The idea would be to study the dynamics very close to the isotropic solution.

First, let us derive a couple of results valid for any gravitational action $S[g_{\mu\nu}]$. It is convenient to define the isotropic expansion rate

$$H(t) := \frac{1}{3}\left(\frac{\dot{a}}{a} + \frac{\dot{b}}{b} + \frac{\dot{c}}{c} \right). \qquad (3.33)$$

The deviation with respect to the isotropic case will be encoded in two functions, $\sigma_1(t)$ and $\sigma_2(t)$, defined implicitly by

$$\frac{\dot{a}}{a} = H + \epsilon_\sigma(2\sigma_1 - \sigma_2), \quad \frac{\dot{b}}{b} = H - \epsilon_\sigma(\sigma_1 - 2\sigma_2), \quad \frac{\dot{c}}{c} = H - \epsilon_\sigma(\sigma_1 + \sigma_2),$$
$$\qquad (3.34)$$

where ϵ_σ is a certain (not necessarily small) fixed parameter controlling the deviation ($\epsilon_\sigma = 0$ for the isotropic case).

The metric (3.32) fulfills the requirements of the *Palais' principle of symmetric criticality* [54] (see also [28, 29, 31]), so one can use the minisuperspace approach and substitute the Ansatz (3.32) in the action and then vary with respect to the free functions $\{\mathcal{N}, a, b, c\}$,

$$\bar{S}[\mathcal{N}, a, b, c] := S[\bar{g}_{\mu\nu}(\mathcal{N}, a, b, c)]. \tag{3.35}$$

It can be shown that, thanks to the Noether identity under diffeomorphisms, the full dynamics is determined by three equations, which can be chosen to be

$$0 = \frac{\delta\bar{S}}{\delta\mathcal{N}}, \qquad \mathrm{E}_1 := \ 0 = \frac{\delta\bar{S}}{\delta c}c - \frac{\delta\bar{S}}{\delta b}b, \qquad \mathrm{E}_2 := \ 0 = \frac{\delta\bar{S}}{\delta c}c - \frac{\delta\bar{S}}{\delta a}a. \tag{3.36}$$

After (and only after!) deriving the equations of motion in the minisuperspace approach, we can take cosmic time $\mathcal{N} = 1$.

3.3.3 Dynamical Equations for the Anisotropies: Strong Coupling Issues

In the theory (3.27), the highest-order derivatives of the anisotropy functions σ_1 and σ_2 appear in the evolution equations $\{\mathrm{E}_1, \mathrm{E}_2\}$, which indeed trivialize in the isotropic case. If we drop the global factor ϵ_σ, they can be written in the following schematic form:

$$\frac{\beta}{M_{\mathrm{Pl}}^2}\left[\epsilon_\sigma\mathbf{M_1}\begin{pmatrix}\dddot{\sigma}_2\\ \dddot{\sigma}_1\end{pmatrix} + \epsilon_\sigma\mathbf{M_2}\begin{pmatrix}\ddot{\sigma}_2\\ \ddot{\sigma}_1\end{pmatrix} + \mathbf{V}\right] + 3M_{\mathrm{Pl}}^2\begin{pmatrix}3H\sigma_2 + \dot{\sigma}_2\\ 3H\sigma_1 + \dot{\sigma}_1\end{pmatrix} = 0. \tag{3.37}$$

where the matrices $\mathbf{M_1}$ and $\mathbf{M_2}$, and the column vector $\mathbf{V}$ start at zeroth order in ϵ_σ.[6] Notice that the higher-order terms containing second and third derivatives of the shear trivialize in the isotropic limit $\epsilon_\sigma \to 0$. Consequently, in this limit, the order of the system of differential equations is abruptly reduced. In other words, we are losing dofs (those associated to the higher-order derivative nature of the Lagrangian) in such a limit.

If we take for instance any Bianchi I metric and expand it around one of the de Sitter solutions (which is isotropic), it can be checked that order by order the resulting equations are of second order. So, around an isotropic background, one cannot recover the exact solution perturbatively. Let us see this more clearly in an example:

---

[6] The components of $\mathbf{M_1}$ depend polynomially on σ_1, σ_2 and H, whereas those of $\mathbf{M_2}$ and $\mathbf{V}$ also depend on $\dot{\sigma}_1, \dot{\sigma}_2$ and the derivatives of H.

Example: Order reduction on specific backgrounds

Consider a differential equation of the type (it should be understood as a schematic expression, since some factors must be introduced to have the appropriate dimensions):

$$0 = E := A(f)\dddot{f} + \ddot{f} + f^2 . \tag{3.38}$$

This equation is of third order in derivatives. Let us now perform an expansion $f = f_0 + f_1 + f_2 + ...$ (we will ignore terms of order greater than 2), where f_0 is a known solution of the theory. The function A can also be split according to the previous expansion, $A = A_0 + A_1 + A_2 + ...$ (notice that $A_0 = A(f_0)$).

If we do the same with the equation $E = E_0 + E_1 + E_2 + ...$, we get

$$E_0 = A_0\dddot{f}_0 + \ddot{f}_0 + f_0^2 , \tag{3.39}$$

$$E_1 = A_0\dddot{f}_1 + A_1\dddot{f}_0 + \ddot{f}_1 + 2f_0f_1 , \tag{3.40}$$

$$E_2 = A_0\dddot{f}_2 + A_1\dddot{f}_1 + A_2\dddot{f}_0 + \ddot{f}_2 + f_1^2 + 2f_0f_2 . \tag{3.41}$$

Here, the equation E_i should be seen as a differential equation for f_i, in which we have substituted f_j with $j < i$ from the previous ones (i.e., we solve order by order). In particular, since f_0 is a solution of the theory the first equation is identically fulfilled. In principle, we see that all the equations continue being of third order for each of the f_i. But this is only true if $A_0 \neq 0$, i.e., if the coefficient in front of the highest-order derivative term in the original equation does not vanish on the background we are considering. In fact, if $A_0 = 0$, the equations for the perturbations become of a lower order (the unknown in each case has been underlined):

$$E_1 = A_1\dddot{f}_0 + \underline{\ddot{f}_1} + 2f_0\underline{f_1} , \tag{3.42}$$

$$E_2 = A_1\dddot{f}_1 + A_2\dddot{f}_0 + \underline{\ddot{f}_2} + f_1^2 + 2f_0\underline{f_2} . \tag{3.43}$$

What we have seen in this example is exactly what happens for CECG around isotropic solutions. This indicates that the perturbative expansion does not capture all of the information about the solution. In more mathematical terms, it can be seen that the isotropic solution belongs to a singular surface in phase space. In fact, in [11] it was shown that it belongs to the intersection of the four singular surfaces of the equation.

The best way to verify that this situation leads to unstable behaviors is by looking at the the evolution of the Hubble parameter $H(t)$ for randomly generated (and small) initial shears σ_1 and σ_2. As shown in Figs. 2 and 3 of [11], the evolution curves of H deviate with respect to the isotropic (unperturbed) cases, clearly indicating that the isotropic background receives important corrections from very small anisotropies.

All the problems mentioned in this section are expected to be found around other backgrounds of these theories. See the recent discussion in the literature regarding spherically symmetric solutions [17, 26, 39]. As it is discussed there (and as we mentioned in Sect. 1.6), these effects have a different interpretation in the EFT framework and are not necessarily problematic.

3.4   Ghosts in Ricci-Based Gravity Theories

In this section we will argue why metric-affine theories that modify GR by adding higher-order curvature corrections are generically plagued by ghosts. More concretely, we will prove that theories where the action is a general function of the inverse metric and the Ricci tensor propagate ghostly dofs unless projective symmetry is imposed, in which case they are GR in disguise [10, 27]. Consider a metric-affine theory described by the action

$$S[g^{\mu\nu}, \Gamma^{\alpha}{}_{\mu\nu}, \psi] = \frac{M_{\mathrm{Pl}}^2}{2} \int \mathrm{d}^D x \sqrt{|g|} F(g^{\mu\nu}, R_{\mu\nu}) + \int \mathrm{d}^D x \sqrt{|g|} \mathcal{L}_{\mathrm{m}}(g^{\mu\nu}, R_{\mu\nu}, \psi)$$

(3.44)

where the Ricci tensor $R_{\mu\nu}$ is that of a general affine connection and ψ stands for a collection of matter fields, which may couple to the connection through the Ricci tensor.[7] This family of theories is known as *(generalized) Ricci-based gravity* theories (gRBGs).

The dynamics for a theory of this family is obtained by varying the action with respect to metric, affine connection and matter fields independently. Particularly, the connection field equations are obtained by setting

$$\int \mathrm{d}^D x \sqrt{|g|} \left(\frac{M_{\mathrm{Pl}}^2}{2} \frac{\partial F}{\partial R_{\mu\nu}} + \frac{\partial \mathcal{L}_{\mathrm{m}}}{\partial R_{\mu\nu}} \right) \delta_\Gamma R_{\mu\nu} \equiv \frac{M_{\mathrm{Pl}}^2}{2} \int \mathrm{d}^D x \sqrt{|q|} q^{\mu\nu} \delta_\Gamma R_{\mu\nu} = 0.$$

(3.45)

This requirement leads to the connection equations

$$\nabla_\lambda \left[\sqrt{|q|} q^{\nu\mu} \right] - \delta^\mu{}_\lambda \nabla_\rho \left[\sqrt{|q|} q^{\nu\rho} \right] = \sqrt{|q|} \left[T^\mu{}_{\lambda\alpha} q^{\nu\alpha} + T^\alpha{}_{\alpha\lambda} q^{\nu\mu} - \delta^\mu{}_\lambda T^\alpha{}_{\alpha\beta} q^{\nu\beta} \right],$$

(3.46)

where ∇ is the covariant derivative of $\Gamma^\alpha{}_{\mu\nu}$. These equations are formally identical to the ones obtained for GR if we substitute the metric for a second rank tensor $q^{\mu\nu}$ with no symmetries. In the case of GR, the solution to this connection equation is given by the Levi-Civita connection for the metric (plus a gauge choice of projective mode). In the case of adding matter that couples to the connection algebraically, an extra term on the right called hypermomentum will arise which will make the solution depart from the Levi-Civita connection of the metric. However, having no symmetry

---

[7] This can be easily generalized to arbitrary couplings with the connection by just introducing a hypermomentum piece in the corresponding equations, see [27].

in the indices of $q^{\mu\nu}$, the above connection equations are formally identical to those of Non Symmetric Gravity [48, 49], a theory that is known to propagate ghostly dofs [23, 24] (see also [6, 7]). In order to benefit from this results, let us introduce the Einstein frame of this theories through the use of field redefinitions and the properties of auxiliary fields.

3.4.1 Einstein Frame for gRBGs and Projective Symmetry

Let us introduce an auxiliary field $\Sigma_{\alpha\beta}$ that allows to linearize the action with respect to the Ricci tensor as

$$S[g, \Gamma, \Sigma, \psi] = \frac{M_{\text{Pl}}{}^2}{2} \int d^D x \sqrt{|g|} \left[F(g^{\mu\nu}, \Sigma_{\mu\nu}) + \frac{2}{M_{\text{Pl}}{}^2} \mathcal{L}_m(g^{\mu\nu}, \Sigma_{\mu\nu}, \psi) \right.$$
$$\left. + \left(\frac{\partial F}{\partial \Sigma_{\mu\nu}} + \frac{2}{M_{\text{Pl}}{}^2} \frac{\partial \mathcal{L}_m}{\partial \Sigma_{\mu\nu}} \right) \left(R_{\mu\nu} - \Sigma_{\mu\nu} \right) \right],$$
(3.47)

where, in some dependencies, indexes are not explicitly written to alleviate the notation. The equivalence between the above action and the original one can be seen by computing the field equations for $\Sigma_{\mu\nu}$, which yield

$$\left[\frac{\partial^2}{\partial \Sigma_{\mu\nu} \partial \Sigma_{\alpha\beta}} \left(F + \frac{2}{M_{\text{Pl}}{}^2} \mathcal{L}_m \right) \right] \left(R_{\mu\nu} - \Sigma_{\mu\nu} \right) = 0 , \qquad (3.48)$$

which are algebraic equation that are uniquely solved by $\Sigma_{\mu\nu} = R_{\mu\nu}$ provided that the second derivative of $F + 2\mathcal{L}_m/M_{\text{Pl}}{}^2$ with respect to $\Sigma_{\mu\nu}$ does not vanish identically. Therefore, $\Sigma_{\mu\nu}$ is indeed auxiliary and by plugging its value in the action (3.47) we recover the original one, thus showing that the above is just a reformulation of the original theory with an extra auxiliary field, and therefore physically equivalent.

Now, we can perform a field redefinition in the new action by defining a new field variable,

$$\sqrt{|q|}q^{\mu\nu} := \sqrt{|g|} \frac{\partial}{\partial \Sigma_{\mu\nu}} \left(F + \frac{2}{M_{\text{Pl}}{}^2} \mathcal{L}_m \right) , \qquad (3.49)$$

and we can replace $\Sigma_{\mu\nu}$ by $q^{\mu\nu}$ by formally inverting this algebraic relation above which yields solutions of the schematic form $\Sigma_{\mu\nu}(q, g, \psi)$. The result is that the action (3.47) now reads

$$S[g, \Gamma, q, \psi] = \frac{M_{\text{Pl}}{}^2}{2} \int d^D x \sqrt{|q|} \left[q^{\mu\nu} R_{\mu\nu} + U(g, q, \psi) \right], \qquad (3.50)$$

with $U(g, q, \psi)$ given by

$$U(g, q, \psi) := \frac{\sqrt{|g|}}{\sqrt{|q|}} \left[F(g, \Sigma) + \frac{2}{M_{\text{Pl}}{}^2} \mathcal{L}_{\text{m}}(g, \Sigma, \psi) \right.$$

$$\left. - \left(\frac{\partial F}{\partial \Sigma_{\mu\nu}} + \frac{2}{M_{\text{Pl}}{}^2} \frac{\partial \mathcal{L}_{\text{m}}}{\partial \Sigma_{\mu\nu}} \right) \Sigma_{\mu\nu} \right] \Bigg|_{\Sigma(q, g, \psi)} . \tag{3.51}$$

From (3.50), we see that the connection equations will be exactly the same as (3.46). At this point, we can also get rid of the metric $g^{\mu\nu}$ because it is also auxiliary, as can be seen by computing its field equations

$$\frac{\partial U}{\partial g^{\mu\nu}} = 0, \tag{3.52}$$

which do not contain derivatives of $g^{\mu\nu}$. Hence, we can solve them to find an algebraic solution for $g^{\mu\nu}(q, \psi)$ in terms of the new object $q^{\mu\nu}$ and the matter fields, and plugging it back into the action we arrive to

$$S[\Gamma, q, \psi] = \frac{M_{\text{Pl}}{}^2}{2} \int \mathrm{d}^D x \sqrt{|q|} q^{\mu\nu} R_{\mu\nu} + \int \mathrm{d}^D x \sqrt{|q|} \mathcal{L}_{\text{m}}^{\text{EF}}(q, \psi), \tag{3.53}$$

where we have defined the Einstein-frame matter action

$$\mathcal{L}_{\text{m}}^{\text{EF}}(q, \psi) := \frac{M_{\text{Pl}}{}^2}{2} U(g, q, \psi) \Bigg|_{g(q, \psi)} . \tag{3.54}$$

We thus see that, with the machinery of field redefinitions and auxiliary fields, we can write the original gRBG theory (3.44) equivalently as (3.53), where the gravitational sector is reduced to a first-order Einstein-Hilbert term with $q^{\mu\nu}$ acting as the (inverse) metric. The corresponding matter action will depend on the original matter fields but with new nonlinear interactions among themselves and $q^{\mu\nu}$.

However, the analogy is not completely satisfied in general because the Ricci tensor of a general affine connection has an antisymmetric piece, and therefore, so does the auxiliary field $\Sigma_{\mu\nu}$ and by extension the object $q^{\mu\nu}$. Hence, in the general case, the Einstein frame of the theory is equivalent to the Non Symmetric Gravity theory formulated by Moffat as explained above, which generically propagates ghostly dofs.

In the case of gRBG theories, there is a safe way of getting rid of these dofs that is implemented by requiring a symmetry in the affine sector so that the pathological dofs will not be reintroduced by quantum corrections (unless there is a gauge anomaly). This symmetry that can ghost-bust gRBG theories is called projective symmetry, and it is realized by having invariance under projective transformations,[8] which in local coordinates are given by

$$\Gamma^{\alpha}{}_{\mu\nu} \overset{\xi}{\longmapsto} \Gamma^{\alpha}{}_{\mu\nu} + \xi_{\mu} \delta^{\alpha}{}_{\nu}, \tag{3.55}$$

---

[8] These transformations do not alter the family of autoparallel curves of the connection (they only introduce a reparameterization of them), hence preserving the affine structure locally introduced in each tangent space by the connection.

where ξ is an arbitrary vector field called projective mode. This transformation of the connection leads to a transformation in the Ricci tensor as

$$R_{\mu\nu} \overset{\xi}{\longmapsto} R_{\mu\nu} - 2\partial_{[\mu}\xi_{\nu]}, \qquad (3.56)$$

so that its symmetric part is invariant but its antisymmetric part changes with the field strength of the projective mode. Hence, we see that imposing projective symmetry in gRBG theories amounts to require that only the symmetric piece of the Ricci tensor appears in the action. If we do so, then the auxiliary field $\Sigma_{\mu\nu}$, and therefore the object $q^{\mu\nu}$, are now both symmetric, and the analogy between the Einstein frame action (3.53) and first-order GR is exactly fulfilled, where $q^{\mu\nu}$ is now a common metric and the connection is given by its Levi-Civita connection plus a spurious projective gauge mode as dictated by the dynamics.

On the other hand, lifting the requirement of projective symmetry allows $q^{\mu\nu}$ to have an antisymmetric part, so that it can be decomposed as

$$\sqrt{|q|}q^{\mu\nu} = \sqrt{|h|}(h^{\mu\nu} + B^{\mu\nu}), \qquad (3.57)$$

where $h^{\mu\nu}$ is symmetric and acts as a usual metric, and $B^{\mu\nu}$ is a 2-form that describes the antisymmetric part of $q^{\mu\nu}$. Thus, beyond the projective mode that will now acquire dynamics due to an explicit breaking of projective symmetry in the action, the 2-form field describing the antisymmetric part of the object $q^{\mu\nu}$ will also introduce new dofs in general. Furthermore, note that the connection equations do not have the Levi-Civita connection of $q^{\mu\nu}$ as a solution anymore, as they will also depend on derivatives of the 2-form $B_{\mu\nu}$. This will end up generating pathological couplings between the 2-form and the curvature of $h_{\mu\nu}$ which lead to the propagation of Ostrogradski instabilities. In the following sections, we will see how these extra dofs are indeed ghostly around arbitrary backgrounds, spoiling the viability of gRBG theories without projective symmetry.

3.4.2 Additional Degrees of Freedom: Ghosts From Splitting the Connection

If we allow the antisymmetric part of the Ricci tensor in the action, projective symmetry is broken and new dofs arise. Let us now show the pathological nature of such dofs by suitably splitting the connection. First, note that for any affine connection $\Gamma^{\alpha}{}_{\mu\nu}$ and symmetric invertible 2nd rank tensor $h^{\mu\nu}$, we can always make the splitting

$$\Gamma^{\alpha}{}_{\mu\nu} = {}^{h}\Gamma^{\alpha}{}_{\mu\nu} + \Upsilon^{\alpha}{}_{\mu\nu}, \qquad (3.58)$$

where $^h\Gamma$ is the Levi-Civita connection of $h^{\mu\nu}$ and $\Upsilon^\alpha{}_{\mu\nu}$ is a rank 3 tensor (see Footnote 1). Then, we can extract the projective mode from $\Upsilon^\alpha{}_{\mu\nu}$ in order to see how it behaves when projective symmetry is explicitly broken:

$$\hat{\Upsilon}^\alpha{}_{\mu\nu} = \Upsilon^\alpha{}_{\mu\nu} + \frac{1}{D-1}\Upsilon_\mu \delta^\alpha{}_\nu, \tag{3.59}$$

where $\Upsilon_\mu := 2\Upsilon^\alpha{}_{[\alpha\mu]}$. The convenience of this choice stems from the fact that $\hat{\Upsilon}^\alpha{}_{\mu\nu}$ satisfies $\hat{\Upsilon}^\alpha{}_{[\alpha\mu]} = 0$ and would allow to solve the connection equations if the 2-form $B_{\mu\nu}$ vanishes. Introducing this splitting explicitly in the action (3.53) we find

$$S[h, B, \hat{\Upsilon}, \Upsilon, \psi] = \frac{M_{\mathrm{Pl}}^2}{2}\int \mathrm{d}^D x \sqrt{|h|}\left[R^h - \frac{2}{D-1}B^{\mu\nu}\partial_{[\mu}\Upsilon_{\nu]} - B^{\mu\nu}\nabla^h_\alpha \hat{\Upsilon}^\alpha{}_{\mu\nu}\right.$$
$$- B^{\mu\nu}\nabla^h_\nu \hat{\Upsilon}^\alpha{}_{\alpha\mu} + \hat{\Upsilon}^\alpha{}_{\alpha\lambda}\hat{\Upsilon}^\lambda{}_\kappa{}^\kappa - \hat{\Upsilon}^{\alpha\mu\lambda}\hat{\Upsilon}_{\lambda\alpha\mu}$$
$$\left. - \hat{\Upsilon}^\alpha{}_{\alpha\lambda}\hat{\Upsilon}^\lambda{}_{\mu\nu}B^{\mu\nu} - \hat{\Upsilon}^\alpha{}_{\nu\lambda}\hat{\Upsilon}^\lambda{}_{\alpha\mu}B^{\mu\nu}\right] + \tilde{S}^{\mathrm{EF}}_{\mathrm{m}}[h, B, \psi], \tag{3.60}$$

where R^h is the curvature scalar of $h_{\mu\nu}$ and $\tilde{S}^{\mathrm{EF}}_{\mathrm{m}}$ is just the matter action after the splitting. From this action, we readily see that the only kinetic term for projective mode Υ_μ occurs through a coupling with $B^{\mu\nu}$, so that around arbitrary $B^{\mu\nu}$ backgrounds, this will render the vector unstable. To make this more explicit, note that the projective mode is oblivious to $\hat{\Upsilon}^\alpha{}_{\mu\nu}$, and let us first consider the relevant sector of the action to describe it around a trivial 2-form background

$$S \supset \int \mathrm{d}^D x \sqrt{|h|}\left[B^{\mu\nu}\partial_{[\mu}\Upsilon_{\nu]} - m^2 B_{\mu\nu}B^{\mu\nu}\right], \tag{3.61}$$

where the mass term will generally be present in $\tilde{S}^{\mathrm{EF}}_{\mathrm{m}}$ due to the old $U(q, g, \psi)$ after solving for $g^{\mu\nu}$ and splitting $q^{\mu\nu} = h^{\mu\nu} + B^{\mu\nu}$, and some factors have been conveniently absorbed into m^2. Now, we can diagonalize the kinetic sector by performing the linear field redefinition

$$B^{\mu\nu} \mapsto \hat{B}^{\mu\nu} + \frac{1}{2m^2}\partial^{[\mu}\Upsilon^{\nu]} \quad \text{and} \quad \Upsilon_\mu \mapsto 2m\Upsilon_\mu, \tag{3.62}$$

which yields

$$S \supset \int \mathrm{d}^D x \sqrt{|h|}\left[\partial^{[\mu}\Upsilon^{\nu]}\partial_{[\mu}\Upsilon_{\nu]} - m^2 \hat{B}_{\mu\nu}\hat{B}^{\mu\nu}\right]. \tag{3.63}$$

This makes apparent that the projective mode is a ghostly vector due to the wrong sign of its kinetic term. One could wonder whether nontrivial backgrounds of $B^{\mu\nu}$ could stabilize the projective mode, acting as a ghost condensate. To see that this cannot be the case, note that around an arbitrary background of $B^{\mu\nu}$, the mass term that would be generated for $B^{\mu\nu}$ will not be diagonal, but rather through a mass matrix $M^{\alpha\beta\mu\nu}$. Then, the relevant piece of the action that describes the dynamics of the projective mode reads

$$S \supset \int \mathrm{d}^D x \sqrt{|h|} \Big[B^{\mu\nu} \partial_{[\mu} \Upsilon_{\nu]} - m^2 M^{\alpha\beta\mu\nu} B_{\mu\nu} B^{\mu\nu} \Big]. \tag{3.64}$$

Hence, the diagonalization of the kinetic sector is carried by the field redefinition

$$B^{\mu\nu} \mapsto \hat{B}^{\mu\nu} + \frac{1}{2m^2} \Lambda^{\mu\nu\alpha\beta} \partial_{[\alpha} \Upsilon_{\beta]} \quad \text{and} \quad \Upsilon_\mu \mapsto 2m\Upsilon_\mu, \tag{3.65}$$

with $\Lambda^{\mu\nu\alpha\beta}$ satisfying $M^{\alpha\beta\rho\sigma} \Lambda_{\rho\sigma}{}^{\mu\nu} = h^{\alpha[\mu} h^{\nu]\beta}$. After these linear field redefinitions, the relevant piece of the action reads

$$S \supset \int \mathrm{d}^D x \sqrt{|h|} \Big[\Lambda^{\alpha\beta\mu\nu} \partial_{[\alpha} \Upsilon_{\beta]} \partial_{[\mu} \Upsilon_{\nu]} - m^2 M^{\alpha\beta\mu\nu} \hat{B}_{\alpha\beta} \hat{B}_{\mu\nu} \Big], \tag{3.66}$$

so that now the possible ghostly character of the vector field is encoded in the eigenvalues of $\Lambda^{\alpha\beta\mu\nu}$. Concretely, the ghosts will be avoided if the eigenvalues of $\Lambda^{\alpha\beta\mu\nu}$ are all negative. For this to be satisfied, $\Lambda^{\alpha\beta\mu\nu}$ needs to have the same signature as $-h^{\alpha[\mu} h^{\nu]\beta}$.[9] On the other hand, the stability of the 2-form mass term requires that the signature of $M^{\alpha\beta\mu\nu}$ is the same than that of $h^{\alpha[\mu} h^{\nu]\beta}$, namely that the mass matrix has positive eigenvalues. However, if the field redefinition diagonalizes the kinetic sector as above, we know that $M^{\alpha\beta\rho\sigma} \Lambda_{\rho\sigma}{}^{\mu\nu} = h^{\alpha[\mu} h^{\nu]\beta}$, so that both conditions cannot be satisfied at the same time, and the instability will appear either as a ghost in the vector sector or as a tachyon in the 2-form sector. Note, as well, that the field redefinition (3.65) looks like a 2-form gauge transformation, so that the gauge invariant kinetic term for the 2-form is not affected and therefore its dynamics have no effect on the propagation of ghosts through the projective mode.

3.4.3 Decoupling Limits and the Stückelberg Trick

In the following section, we will analyze the presence of ghostly dofs in the decoupling limit for the 2-form field. Before that, let us clarify what we mean by decoupling limit in this short section. A *decoupling limit* for a set of dofs is a limit taken in the parameters of the theory (masses, scales or couplings) in which these dofs decouple from the rest, so that some properties of the full theory may be easier to unveil. Note, however, that a decoupling limit represents only a certain regime of the theory, so some properties derived in such a limit will not generalize to the full theory. For instance, being ghost-free in a decoupling limit does not mean that the theory is fully ghost-free, although finding ghosts in the decoupling limit typically implies that the ghosts will be in the full theory as well.

---

[9] Note that $h^{\mu\nu}$ is a Lorentzian metric so that $h^{\alpha[\mu} h^{\nu]\beta}$ has positive eigenvalues.

Decoupling limits are trivially implemented in some cases. For instance, if we take the limit of zero fine-structure constant in Maxwell electrodynamics with charged fermions, one obtains a theory of a bunch of free fermions and a free massless vector field where all the dofs are decoupled. However, in some other cases, taking this limit in a physically sensible manner is not so trivial, as a naive limit may alter the number of dofs of the theory. This happens, for example, if one considers the massless limit of Proca theory, in which one dof is lost. In these cases one has to be more careful in taking the decoupling limit of the theory.

These problems appear when some gauge symmetries are recovered in such a limit, since they prevent some dofs from propagating. In these cases, a way of taking the decoupling limit is by introducing Stückelberg fields to restore the gauge symmetry in the full theory and then take their decoupling limit, which leads to the decoupling of the Stückelberg modes. This is sometimes referred to as *Stückelberg trick*, and the result is a theory with a gauge field decoupled from the Stückelberg modes, which encode the extra dofs that are present of the full theory. Let us write down an explicit example on how to implement the Stückelberg trick for a Proca theory.

Example: Stückelberg field for the Proca theory

Start from the Lagrangian

$$\mathcal{L}_{\text{Proca}} = -\frac{1}{4} F_{\mu\nu} F^{\mu\nu} + \frac{1}{2} m^2 A_\mu A^\mu. \tag{3.67}$$

The mass term breaks the U(1) gauge symmetry of the kinetic term since the transformation $A_\mu \mapsto \hat{A}_\mu + \partial_\mu \varphi$ leads to the addition of the terms $m^2 \hat{A}^\mu \partial_\mu \varphi$ and $(m^2/2) \partial_\mu \varphi \partial^\mu \varphi$ in the Lagrangian, which cannot be written as a total derivative. The Stückelberg trick here consists on two steps. 1) first restore the U(1) gauge-invariance of the vector field by introducing the Stückelberg field φ by making the replacement $A_\mu \mapsto \hat{A}_\mu + m^{-1} \partial_\mu \varphi$, which leads to the Stückelberg field Lagrangian (2.67) after a linear field redefinition $\varphi \mapsto m^{-1} \varphi$, namely

$$\mathcal{L}_{\text{Stück}} = -\frac{1}{4} \hat{F}_{\mu\nu} \hat{F}^{\mu\nu} + \frac{1}{2} m^2 \hat{A}_\mu \hat{A}^\mu + m^2 \hat{A}^\mu \partial_\mu \varphi + \frac{1}{2} \partial^\mu \varphi \partial_\mu \varphi. \tag{3.68}$$

Here we see that the Stückelberg field is coupled to the massive vector through the vector's mass. In step 2) we consider the massless limit for the vector field, which is the decoupling limit for the Stückelberg mode, and yields

$$\mathcal{L}_{\text{Stück}} = -\frac{1}{4} \hat{F}_{\mu\nu} \hat{F}^{\mu\nu} + \frac{1}{2} \partial^\mu \varphi \partial_\mu \varphi, \tag{3.69}$$

which indeed describes three dofs as the original theory, two encoded in the massless vector and one encoded in the Stückelberg field. This could be heuristically associated to the longitudinal polarization that is lost in the massless limit of the Proca theory.

This trick can be generalized to more general gauge theories, and we will use it in the next section to show how the the 2-form field $B_{\mu\nu}$ and the projective mode propagate unstable dofs.

3.4.4 Ghosts in the Decoupling Limit for the 2-Form

First, consider the 2-form sector perturbatively, so that from (3.53) we find at quadratic order[10]

$$
S^{(2)} = \int d^D x \sqrt{|h|} \left[\frac{M_{\mathrm{Pl}}^2}{2} R^h - \frac{1}{12} H_{\mu\nu\rho} H^{\mu\nu\rho} - \frac{1}{4} m^2 B^{\mu\nu} B_{\mu\nu} - \frac{\sqrt{2} M_{\mathrm{Pl}}}{3} B^{\mu\nu} \partial_{[\mu} \Upsilon_{\nu]} \right.
$$
$$
\left. + \frac{1}{4} R^h B^{\mu\nu} B_{\mu\nu} - R^h_{\mu\nu\alpha\beta} B^{\mu\alpha} B^{\nu\beta} \right], \tag{3.70}
$$

where $H_{\mu\alpha\beta} := \partial_{[\mu} B_{\alpha\beta]}$ is the field strength of the 2-form, which provides a ghost-free (and gauge invariant) kinetic term for a 2-form. The mass of the 2-form is generated by the Einstein frame matter Lagrangian in general. The Stückelberg trick for the 2-form is parallel to that for the vector field, namely, introduce Stückelberg fields b_ν for the 2-form by the replacement

$$
B_{\mu\nu} \mapsto \hat{B}_{\mu\nu} + \frac{2}{m} \partial_{[\mu} b_{\nu]}. \tag{3.71}
$$

After introducing the Stückelberg field, performing the field redefinition

$$
\Upsilon_\mu \mapsto \frac{3m}{\sqrt{2} M_{\mathrm{Pl}}} \hat{\Upsilon}_\mu, \tag{3.72}
$$

and taking the flat spacetime limit, we arrive to the action

$$
S^{(2)} = \int d^D x \left(-\frac{1}{12} \hat{H}_{\mu\nu\rho} \hat{H}^{\mu\nu\rho} - \frac{1}{4} m^2 \hat{B}^{\mu\nu} \hat{B}_{\mu\nu} - \partial^{[\mu} b^{\nu]} \partial_{[\mu} b_{\nu]} - m \hat{B}^{\mu\nu} \partial_{[\mu} b_{\nu]} \right.
$$
$$
\left. - m \hat{B}^{\mu\nu} \partial_{[\mu} \hat{\Upsilon}_{\nu]} - 2 \partial^{[\mu} b^{\nu]} \partial_{[\mu} \hat{\Upsilon}_{\nu]} \right). \tag{3.73}
$$

We can now take the decoupling limit for the Stückelberg as $m \to 0$, which yields

$$
S^{(2)} = \int d^D x \left(-\frac{1}{12} \hat{H}_{\mu\nu\rho} \hat{H}^{\mu\nu\rho} - \partial^{[\mu} b^{\nu]} \partial_{[\mu} b_{\nu]} - 2 \partial^{[\mu} b^{\nu]} \partial_{[\mu} \hat{\Upsilon}_{\nu]} \right), \tag{3.74}
$$

---

[10] In 4 dimensions and after a redefinition $B_{\mu\nu} \mapsto 2 B_{\mu\nu}/M_{\mathrm{Pl}}^2$.

which describes the same number of dofs than a massive 2-form plus a massless vector, as the original action. Now, we can write the kinetic terms for the vectors as

$$
\left(\partial_{[\mu} b_{\nu]}\ \partial_{[\mu}\hat{\Upsilon}_{\nu]}\right)\begin{pmatrix} -1 & -1 \\ -1 & 0 \end{pmatrix}\begin{pmatrix} \partial^{[\mu} b^{\nu]} \\ \partial^{[\mu}\hat{\Upsilon}^{\nu]} \end{pmatrix}. \tag{3.75}
$$

Therefore, we see that the kinetic matrix for the vector sector has eigenvalues $(-1 \pm \sqrt{5})/2$. Since one is negative, this signals the presence of a ghostly dof, either in the Stückelberg field or in the projective mode. This can also be explicitly seen by diagonalizing the vectorial sector through the linear field redefinition

$$
b_\mu \mapsto A_\mu + \xi_\mu \qquad \text{and} \qquad \Gamma_\mu \mapsto -2\xi_\mu, \tag{3.76}
$$

after which the second order action once the decoupling limit has been taken reads

$$
S^{(2)} = \int \mathrm{d}^D x \left(-\frac{1}{12}\hat{H}_{\mu\nu\rho}\hat{H}^{\mu\nu\rho} - \partial^{[\mu} A^{\nu]}\partial_{[\mu} A_{\nu]} + \partial^{[\mu}\xi^{\nu]}\partial_{[\mu}\xi_{\nu]} \right), \tag{3.77}
$$

where it can be seen that ξ is a ghost around Minkowski spacetime. This will prevail for arbitrary backgrounds of $h^{\mu\nu}$ where, apart from the ghostly projective mode, we will also have problems in the 2-form sector due to its couplings to the curvature of the symmetric metric (see Eq. (3.70)), which excite Ostrogradski instabilities through the appearance of non-degenerate 2nd-order derivatives of $h^{\mu\nu}$ in the action which will manifest as further ghosts in the 2-form sector. There are some technical details in this analysis that have been overlooked, such as the possibility of doing a quadratic field redefinition of the symmetric and antisymmetric piece of the metric. As well, interestingly, there are ways of getting rid of these pathologies within gRBG theories that relay on placing geometric constraints, rather than symmetry requirements. For a more detailed account of the instabilities that arise in gRBG without projective symmetry, as well as possible evasion mechanisms, or the impact of the analysis for more general metric-affine theories, the reader is referred to [7, 27].

Appendix A
List of Problems

Problems

Conventions: we choose the signature $\eta_{\mu\nu} = \mathrm{diag}(+1, -1, -1, -1)$.

A.1 Bounded and Unbounded Growth

(a) Show that single-mode perturbations of a free field in Minkowski spacetime are bounded, namely, that for any initial perturbation φ_k to a mode $e^{i(\mathbf{k}\cdot x - \omega_k t)}$, there exists a constant C such that $|\varphi_k| < C$ at all times.

(b) Show also that, for a system with a gradient instability, given an initial amplitude A, and an arbitrarily small time interval Δt, there are modes with unbounded growth within that time interval. Show also that is not true for a system with a tachyonic instability.

A.2 Jeans' Instability

For a self gravitating cloud of dust of density ρ and pressure P, the mass and momentum conservation equations are

$$\partial_t \rho + \partial_i(\rho u^i) = 0 \quad \text{and} \quad \partial_t u^j + u^i \partial_i u^j + \rho^{-1}\partial^j P + \partial^j \Phi = 0$$

where u^i is the velocity vector of a fluid element, Φ is the gravitational potential satisfying the Poisson equation $\Delta \Phi = \kappa \rho$, and indices are raised and lowered with δ_{ij}. Assuming a steady initial state $\bar{u}^i = 0$ with constant density and pressure $\bar{\rho}$ and $\bar{P}$:

(a) Derive the (2nd order) equation followed by linear adiabatic perturbations $\delta P = c_s^2 \delta\rho$.

(b) For the case where gravitation is turned off, $\kappa = 0$, describe the behavior of adiabatic density perturbations. Is the initial configuration of the system stable?

(c) Repeat the analysis taking into account the gravitational interaction. What qualitative differences do you find and what is the physical interpretation for them?

© The Editor(s) (if applicable) and The Author(s), under exclusive license to Springer Nature Switzerland AG 2023
A. Delhom et al., *Instabilities in Field Theory*, SpringerBriefs in Physics, https://doi.org/10.1007/978-3-031-40433-7

(d) If one takes into account the expansion of the universe, and define the fractional density perturbation $\delta := \delta\rho/\bar{\rho}$, the equation that you have found becomes

$$\partial_t^2 \delta + 2H\partial_t \delta - \frac{c_s^2}{a^2} \Delta\delta - \kappa\bar{\rho}\delta = 0, \tag{A.1}$$

where a is the scale factor ans H the Hubble rate. For constant positive expansion rate $H > 0$, compute the general solution to this equation. Discuss what is the role of the expansion rate and whether it changes anything about the stability of the perturbations from a qualitative point of view.

A.3 Non-canonical Kinetic and Mass Terms: Uncovering Instabilities Through Field Redefinitions

The following Lagrangian density describes a field theory with two fields ϕ and ψ propagating around a particular background defined by the coefficients a, b, c and d.

$$\mathcal{L} = a\,\partial^\mu\phi\partial_\mu\phi + b\,\partial^\mu\psi\partial_\mu\psi + c\,\partial^\mu\phi\partial_\mu\psi + d\,\phi\psi$$

(a) find the kinetic, gradient and mass matrices for the theory.
(b) For $b = 0$ (and non-vanishing a, c), show that there is a ghost degree of freedom.
(c) Show that whenever $4ab \neq c$ the theory propagates 2 degrees of freedom. Derive the condition/s that must be satisfied by the coefficients a, b, c for the theory to be free of ghosts and/or gradient instabilities.
(d) What relations must the coefficients satisfy in order for the background to potentially suffer from a strong coupling instability? If this condition is satisfied for this background, in what cases will we have this instabilities?
(e) Find the field redefinition that diagonalizes the mass matrix (i.e. the fields that are mass eigenstates) and show that there is always a tachyonic instability unless the fields are massless.

A.4 Ostrogradski Procedure

The following Lagrangian density describes a field theory with two fields φ and ψ,

$$\mathcal{L} = \frac{1}{2}(a\partial_\mu\partial_\nu\varphi\partial^\mu\partial^\nu\varphi + \partial_\mu\psi\partial^\mu\psi + b\partial_\mu\psi\partial^\mu\varphi + b\psi\Box\varphi)\,\mathrm{d}^3x\,, \tag{A.2}$$

where $\Box := \eta^{\mu\nu}\partial_\mu\partial_\nu$.

(a) Is it possible to simplify the Lagrangian?
(b) The Hessian matrix is given by:

$$\begin{pmatrix} \dfrac{\partial^2 \mathcal{L}}{\partial\ddot{\varphi}^2} & \dfrac{\partial^2 \mathcal{L}}{\partial\dot{\psi}\partial\ddot{\varphi}} \\[3mm] \dfrac{\partial^2 \mathcal{L}}{\partial\ddot{\varphi}\partial\dot{\psi}} & \dfrac{\partial^2 \mathcal{L}}{\partial\dot{\psi}^2} \end{pmatrix} \tag{A.3}$$

Calculate it and show that whenever $a \neq 0$, the theory propagates more than 2 degrees of freedom.

(c) Under the condition $b = 0$, compute the Ostrogradski Hamiltonian. Is the Ostrogradski theorem fulfilled in this case? Why?

A.5 Field Redefinitions and Detection of Ghosts

Consider a theory depending on $\{\Psi^A, A_\mu, B_\nu\}$, where $\{\Psi^A\}$ is a family of arbitrary fields (vectors, scalars, tensors...) that propagate healthily as dictated by the Lagrangian

$$\mathcal{L} = a\, F_{\mu\nu}(A) F^{\mu\nu}(B) + (\text{kinetic terms for } \Psi^A) + V(\Psi^A, A_\mu, B_\nu) \qquad (A.4)$$

where $F_{\mu\nu}(X) := \partial_\mu X_\nu - \partial_\nu X_\mu$, a is a real number and V does not depend on the derivatives of the fields.

(a) Diagonalize the kinetic sector of A_μ and B_μ by doing the field redefinition,

$$A_\mu = U_\mu + V_\mu \qquad B_\mu = U_\mu - V_\mu , \qquad (A.5)$$

and show that the presence of the first term (i.e., $a \neq 0$) automatically indicates that one of the vectors is a ghost.

Hint: Remember that the canonical normalization for a vector is $-\frac{1}{4} F_{\mu\nu} F^{\mu\nu}$.

(b) Indeed it is not necessary to perform the previous field redefinition. The fact that there is a ghost can be directly seen from the kinetic matrix (even if it has not been diagonalized). To see this, rewrite the kinetic sector of A_μ and B_μ as

$$a\, F_{\mu\nu}(A) F^{\mu\nu}(B) = -\frac{1}{4} F_{\mu\nu}(X^i)\, M_{ij}\, F^{\mu\nu}(X^j) \qquad (A.6)$$

with $X^i = (A, B)$, and calculate the determinant of the kinetic matrix M_{ij}. What can we conclude from it?

(c) Assume that there are three vectors (instead of two) whose kinetic sector is separated from the rest of the theory. What can be said about the presence of ghosts depending on the sign of the determinant of the kinetic matrix?

A.6 Dangerous Interaction Terms Between Vectors

Consider a theory with two interacting massive vector fields $\{A_\mu, B_\mu\}$ with (a priori) safe kinetic sector:

$$\mathcal{L} = -\frac{1}{4} F_{\mu\nu}(A) F^{\mu\nu}(A) - \frac{1}{4} F_{\mu\nu}(B) F^{\mu\nu}(B) + \frac{1}{2} m_A^2 A_\mu A^\mu + \frac{1}{2} m_B^2 B_\mu B^\mu + \mathcal{L}_{\text{int}} . \qquad (A.7)$$

Show that a self-interaction of the type

$$\mathcal{L}_{\text{int}}^{(1)} = A_\mu A^\mu \partial_\nu A^\nu , \qquad (A.8)$$

for one (or both) vectors is a healthy interaction, whereas the interactions that mix the vectors,

$$\mathcal{L}_{int}^{(2)} = A_\mu A^\mu \partial_\nu B^\nu \quad \text{and} \quad \mathcal{L}_{int}^{(3)} = A_\mu B^\mu \partial_\nu A^\nu, \tag{A.9}$$

introduce a ghost.

Hint: First extract the longitudinal part of each of the vectors, i.e., perform $X_\mu \to \partial_\mu \phi + Y_\mu$, where the parts Y_μ are transversal ($\partial_\mu Y^\mu = 0$), and analyze the higher-order terms (in time derivatives) for the longitudinal parts ϕ's. Then, study the contribution of these longitudinal parts to the Hessian.

A.7 Theory with Two Scalars and Second-Order Derivatives

Consider the Lagrangian

$$\mathcal{L} = \frac{1}{2}\ddot{\phi}^2 + \frac{1}{2}\ddot{\psi}^2 + \ddot{\phi}\ddot{\psi} - \frac{1}{2}m^2(\phi^2 + \psi^2) \quad m^2 > 0. \tag{A.10}$$

How many degrees of freedom does this theory propagate? Is it ghost-free?

Hint: Check first the Hessian, and then try to find a field redefinition that simplifies the higher-order part.

Solutions

A.1 Bounded and Unbounded Growth

(a) This is already outlined in the first example of Sect. 1.2. A free field in Minkowski spacetime satisfies linear equations $\mathcal{E}(\phi) = 0$. That means that the space of solutions is a vector space, since the sum of two solutions is also a solution (you can check all the properties of a vector space hold). Hence, if we have an exact solution ϕ_0, due to linearity,

$$\mathcal{E}(\phi_0 + \varphi) = \mathcal{E}(\phi_0) + \mathcal{E}(\varphi), \tag{A.11}$$

perturbations φ will also exactly satisfy $\mathcal{E}(\varphi) = 0$. Therefore, in this case, perturbations around exact solution for the scalar field will also be exact solutions. Due to Poincaré invariance, the spacetime dependence of the solutions will be given by plane waves $e^{\pm i(\mathbf{k}\cdot\mathbf{x}-\omega t)}$. Hence, the perturbations will be of the form

$$\varphi_{\mathbf{k}}(t, \mathbf{x}) = A_{\mathbf{k}} e^{i(\mathbf{k}\cdot\mathbf{x}-\omega t)} \tag{A.12}$$

where the initial amplitude of the wave profile $A_{\mathbf{k}}$ is constant in spacetime (i.e., it only depends on the mode $\mathbf{k}$) and where

$$\omega = \sqrt{|\mathbf{k}|^2 + m^2} \tag{A.13}$$

for the above to be a solution. Hence, any constant $C > |A_{\mathbf{k}}|$ bounds the perturbations at all times. The fact that functions of the above form (A.12) are exact solutions is general for any free relativistic theory with Poincaré invariance, and is due precisely to homogeneity of the background. For different types of fields, the associated $A_{\mathbf{k}}$ will take values in the corresponding irreducible representation of the Poincaré group (i.e. a scalar, polarization vector, spinor, etc.).

(b) A system with a gradient instability develops imaginary frequencies for modes with high enough momenta. Solutions with high enough momenta of given amplitude A and after a time interval Δt will be of the form $Ae^{\pm \omega \Delta t - i\mathbf{k}\cdot\mathbf{x}}$. If there was a constant C that bounds the growth of all modes, it would have to satisfy

$$\left|Ae^{\pm \omega \Delta t - i\mathbf{k}\cdot\mathbf{x}}\right| = |A|e^{\pm|\omega|\Delta t} < C. \tag{A.14}$$

For the negative sign, since $|\omega| > 0$, the modes are decaying in time, and it is enough to have $|A| < C$ and the requirement will be satisfied for all $\Delta t > 0$. However, for the positive sign, the modes are exponential growing in time. That means that for any $A > 0$ and time interval $\Delta t > 0$, if there existed a constant $C > 0$ bounding the growth of all modes in such time interval, it would need to satisfy

$$|\omega| < \frac{\log|C/A|}{\Delta t}. \tag{A.15}$$

Since gradient instabilities occur for arbitrarily high $|\omega|$, for $A > 0$ and $\Delta t > 0$ there is no $C < \infty$ such that the above requirement holds for all modes, proving the first statement. For the tachyonic case, on the other hand, the exponential growth occurs only for $|\omega| < \omega_c$. Hence, any real constant

$$C(A, \Delta t) > Ae^{\omega_c \Delta t} \tag{A.16}$$

is a bound to all tachyonic modes with initial amplitude A after a time interval not greater than Δt.

A.2 Jeans' Instability

(a) A perturbed fluid will be described by $u^i = \bar{u}^i + \delta u^i$, $\rho = \bar{\rho} + \delta\rho$ and $P = \bar{P} + \delta P$. Assuming that the background quantities satisfy the mass and momentum conservation equations, we find the following equations for linear perturbations

$$\partial_t \delta\rho + \partial_i(\delta\rho\bar{u}^i + \bar{\rho}\delta u^i) = 0 \tag{A.17}$$

$$\partial_t \delta u^j + \delta u^i \partial_i \bar{u}^j + \bar{u}^i \partial_i \delta u^j - \bar{\rho}^{-2}(\partial^j \bar{P})\delta\rho + \bar{\rho}^{-1}\partial^j \delta P + \partial^j \delta\Phi = 0 \tag{A.18}$$

We have a background with vanishing velocity $\bar{u}^i = 0$ and with homogeneous and time-independent density and pressure profiles, so that we can neglect their derivatives. Under these assumptions, by taking the divergence of the second equation, using the result in the time derivative of the first equation, and using also $\Delta\delta\Phi = \kappa\delta\rho$, we find the following second order PDE for the linear perturbations

$$\partial_t^2\delta\rho - \Delta\delta P = \kappa\bar{\rho}\delta\rho. \tag{A.19}$$

Adiabatic perturbations will therefore satisfy

$$(\partial_t^2 - c_s^2\nabla^2 - \kappa\bar{\rho})\delta\rho = 0 \tag{A.20}$$

(b) If gravity is turned off, this is a KG equation for a massless scalar field with a Lorentz symmetry given by the soundcones (instead of lightcones). Hence, the general solution is of the form (1.3) where the dispersion relation is

$$\omega = c_s|\mathbf{k}|. \tag{A.21}$$

Adiabatic perturbations therefore evolve as perturbations on top of Minkowski space, which are stable as discussed above.

(c) If we now take into account the gravitational interaction, we see that it induces a negative mass squared term $\mu = -\kappa\bar{\rho}$ which implies that the background considered has a tachyonic instability. The dispersion relation in this case will be

$$\omega = \sqrt{c_s^2|\mathbf{k}|^2 - \kappa\bar{\rho}}, \tag{A.22}$$

so that modes satisfying

$$|\mathbf{k}| < k_J := \frac{\sqrt{\kappa\bar{\rho}}}{c_s} \tag{A.23}$$

will exhibit tachyonic growth. The time scale of the instability is dictated by the $\mathbf{k} \to 0$ mode, and it is $T_J = 1/\sqrt{\kappa\bar{\rho}}$. The physical interpretation goes as follows. A a homogeneous cloud of gas is unstable due to self-gravitational interaction. No matter how small its density is, there is a length scale, the Jeans' length $\lambda_J = 2\pi k_J^{-1}$, such that perturbations with higher wavelength make the density grow exponentially. Hence, homogeneous gas clouds are unstable, and doomed to collapse gravitationally.

(d) Now, the background quantities will change in time. Assuming that they vary slow enough, the general solution to the PDE is also a superposition of complex exponentials $e^{-i(\mathbf{k}\cdot\mathbf{x}-\omega t)}$ where the dispersion relation is modified to

$$\omega = i\frac{3H}{2} \pm \sqrt{c_s^2(t)|\mathbf{k}|^2 - \left(\kappa\bar{\rho} + \frac{9H^2}{4}\right)}, \tag{A.24}$$

where $c_s(t) = c_s/a(t)$. This implies that there is a decaying exponential factor $e^{-3Ht/2}$ that modulates the perturbations and that the scale controlling the instability becomes time-dependent. The decaying mode in the no-expansion case now decays faster. On the other hand, the growing mode is now governed by the exponent

$$\alpha = \left(-\frac{3H}{2} + \sqrt{\frac{9H^2}{4} + \kappa \bar{\rho}} \right) t. \tag{A.25}$$

Let us analyze the limits of no expansion and of negligible gravitational interaction compared to the expansion (or infinitely fast expansion). These are respectively given by

$$\alpha = \sqrt{\kappa \bar{\rho}} - \frac{3H}{2} + O\left(H^2\right), \tag{A.26}$$

$$\alpha = \frac{\kappa \bar{\rho}}{3H} + O\left(H^{-2}\right). \tag{A.27}$$

We can see how in the limit of no expansion, we recover the previous result for the Jeans' length. On the other hand, in the limit in which the gravitational interaction is negligible with respect to the expansion, the expansion is able to dilute the perturbations quickly enough so that the tachyonic growth never occurs.

A.3 Non-canonical Kinetic and Mass Terms: Uncovering Instabilities Through Field Redefinitions

(a) If we perform the 3+1 splitting of the derivatives, we find:

$$\mathcal{L} = a\dot{\phi}^2 + b\dot{\psi}^2 + c\dot{\phi}\dot{\psi} - a\partial_i\phi\partial^i\phi - b\partial_i\psi\partial^i\psi - c\partial_i\phi\partial^i\psi + d\phi\psi. \tag{A.28}$$

where the dot denotes time derivative and spatial (Latin) indices are raised with the Euclidean metric δ^{ij}. From this expression one can read the kinetic, gradient and mass matrices in the basis $\{\varphi^I\} = \{\phi, \psi\}$, which are respectively:

$$(a_{IJ}) = \begin{pmatrix} 2a & c \\ c & 2b \end{pmatrix}, \quad (b_{IJ}) = \begin{pmatrix} 2a & c \\ c & 2b \end{pmatrix}, \quad (\mu_{IJ}) = \begin{pmatrix} 0 & -d \\ -d & 0 \end{pmatrix}. \tag{A.29}$$

(b) For $b = 0$ and non-vanishing a, c, $\det(a_{IJ}) < 0$ and, for a two-dimensional symmetric matrix, this implies that the two eigenvalues have opposite signs. The one that is negative correspond to the ghostly degree of freedom of the theory.

(c) If $\det(a_{IJ}) = 4ab - c^3 \neq 0$, both eigenvalues of the kinetic matrix are non-vanishing. Therefore, the two fields (each of them carrying one degree of freedom) are propagating.

Since the kinetic and the gradient matrices coincide, the theory is free of gradient instabilities. Ghosts are however possible if either of the eigenvalues of a_{IJ} is negative, i.e., we need it to be positive definite. A necessary and sufficient condition for this is having positive leading principal minors (these are the determinants with upper-left corner nailed to the $(1, 1)$ element of the matrix). In our case, this means $a > 0$ and $4ab - c^2 > 0$.

(d) The strong coupling appears whenever some degrees of freedom are not active at leading order around that background. Having a non-propagating degree of freedom is equivalent to having a null eigenvalue for the kinetic matrix. Therefore, if $\det(a_{IJ}) = 4ab - c^3 = 0$ there is a potential strong coupling problem. In general, having a strong coupling problem will lead to unstable behaviours (except if the theory is treated appropriately, e.g. as an EFT). This depends on the particular theory and should be studied carefully.

(e) If we perform $\phi = (\alpha + \beta)/\sqrt{2}$ and $\psi = (\alpha - \beta)/\sqrt{2}$, the mass term becomes:

$$d\phi\psi = \frac{d}{2}(\alpha^2 - \beta^2). \tag{A.30}$$

Both mass terms have been decoupled. However we notice that for any non-vanishing real value of d both mass terms have opposite signs, so at least one of them is tachyonic. This is avoided only for $d = 0$, i.e. if both fields are massless.

A.4 Ostrogradski Procedure

(a) Yes, as proven below using the Leibniz rule, the last two terms constitute a boundary term so for dynamical purposes (ignoring boundary effects) they can be omitted:

$$b\partial_\mu\psi\partial^\mu\varphi + b\psi\Box\varphi = b\partial_\mu(\psi\partial^\mu\varphi). \tag{A.31}$$

(b) Calculating the Hessian matrix as indicated we find:

$$\mathcal{K} = \begin{pmatrix} a & 0 \\ 0 & 1 \end{pmatrix}, \tag{A.32}$$

where we have used the simplification found in the previous question (which tells us that the two fields are decoupled).

Since the Lagrangian is composed of 2 fields and contains second-order time derivatives, if the Hessian is not degenerate (non-zero determinant) the theory would propagate more than two degrees of freedom. We calculate the determinant of the Hessian and obtain

$$\det\mathcal{K} = a, \tag{A.33}$$

which clearly is non-zero if $a \neq 0$.

(c) Let us re-write the Lagrangian as

$$\mathcal{L} = \frac{1}{2}(a\ddot{\varphi}^2 + \dot{\psi}^2 + X), \tag{A.34}$$

where X represents all the terms without time derivatives. The sector depending on ψ follows exactly the standard Hamiltonian procedure:

$$q_3 := \psi, \qquad \pi_3 = \frac{\delta L}{\delta \dot{\psi}} = \dot{\psi} \tag{A.35}$$

For the other one we introduce the Ostrogradski new variable:

$$q_1 := \varphi, \qquad \pi_1 := \frac{\delta L}{\delta \dot{\varphi}} - \frac{d}{dt}\frac{\delta L}{\delta \ddot{\varphi}} = 2a\Delta\dot{\varphi} - a\dddot{\varphi},$$

$$q_2 := \dot{\varphi}, \qquad \pi_2 := \frac{\delta L}{\delta \dddot{\varphi}} = a\ddot{\varphi}, \tag{A.36}$$

where Δ is the Laplacian. Then

$$\ddot{\varphi} = F(\pi_2) = \frac{1}{a}\pi_2, \tag{A.37}$$

and the Hamiltonian density is:

$$\begin{aligned}
\mathcal{H} &= \dot{q}_1\pi_1 + \dot{q}_2\pi_2 + \dot{q}_3\pi_3 - \mathcal{L} \\
&= q_2\pi_1 + \frac{1}{a}(\pi_2)^2 + (\pi_3)^2 - \frac{1}{2}(a\ddot{\varphi}^2 + \dot{\psi}^2 + X) \\
&= q_2\pi_1 + \frac{1}{a}(\pi_2)^2 + (\pi_3)^2 - \frac{1}{2a}(\pi_2)^2 - \frac{1}{2}(\pi_3)^2 - \frac{1}{2}X \\
&= q_2\pi_1 + \frac{1}{2a}(\pi_2)^2 + \frac{1}{2}(\pi_3)^2 - \frac{1}{2}X .
\end{aligned} \tag{A.38}$$

Now we express X in terms of the new variables:

$$\begin{aligned}
X &= -2a\partial_i\dot{\varphi}\partial^i\dot{\varphi} + a\partial_i\partial_j\varphi\partial^i\partial^j\varphi - \partial_i\psi\partial^j\psi \\
&= -2a|\nabla q_2|^2 + a(\Delta q_1)^2 - |\nabla q_3|^2 ,
\end{aligned} \tag{A.39}$$

where indices are raised with the Euclidean metric δ_{ij}. So finally,

$$H = \int d^3x \frac{1}{2}\left[2q_2\pi_1 + \frac{1}{a}(\pi_2)^2 + (\pi_3)^2 - a(\Delta q_1)^2 + 2a|\nabla q_2|^2 + |\nabla q_3|^2\right]. \tag{A.40}$$

We see that the Hamiltonian is linear in π_1, in agreement with the Ostrogradski theorem. The theorem holds because the condition obtained in a) reflects that the Lagrangian is non-degenerate.

A.5 Field Redefinitions and Detection of Ghosts

(a) Under the given redefinition, since $F_{\mu\nu}(A)$ is linear in A, we can schematically show:

$$F(A)F(B) = F(U)F(U) - F(V)F(V). \tag{A.41}$$

Therefore, if one of them is canonically normalized, the other one has the wrong sign in front. Hence, there is necessarily a ghost.

(b) The kinetic matrix is:

$$(M_{ij}) = \begin{pmatrix} 0 & -4 \\ -4 & 0 \end{pmatrix}, \tag{A.42}$$

which has a negative determinant. Since the matrix is 2-dimensional, this means that one of the eigenvalues is positive and the other negative, hence indicating that one of the fields is a ghost and the other one propagates safely.

(c) We can consider 3 cases depending on the value of the determinant:

- If the determinant is zero, it means that at least one of the three vectors do not propagate. In principle, from the whole determinant one cannot conclude anything about the rest. Further analysis is needed.
- If the determinant is positive there are two possibilities: either the 3 are safe vectors or 2 of them are ghosts.
- If the determinant is negative there are two possibilities: either the 3 are ghostly or only 1 of them is a ghost.

Observe that, generically, a negative determinant indicates that there is at least one ghost.

A.6 Dangerous Interaction Terms Between Vectors

Let us call ϕ and ψ the longitudinal parts of A_μ and B_μ, respectively.

For the first case we can extract the longitudinal part as

$$\mathcal{L}_{\text{int}}^{(1)} = A_\mu A^\mu \partial_\nu A^\nu \quad \rightarrow \quad \partial_\mu \phi \partial^\mu \phi \Box \phi + \ldots = \dot{\phi}^2 \ddot{\phi} - (\delta^{ij} \partial_i \phi \partial_j \phi) \ddot{\phi} + \ldots, \tag{A.43}$$

where we just show explicitly the higher-order terms in time derivatives. Analyzing the previous expression, we realize the only term that could potentially introduce more than second-order time derivatives in the field equations is the first one. Nevertheless, as seen below, such a term can be rewritten as a total derivative, and hence it does not contribute to the field equations:

$$\dot{\phi}^2 \ddot{\phi} = \partial_0 \left(\frac{1}{3} \dot{\phi}^3 \right). \tag{A.44}$$

On the other hand, for the other two interactions, we find

$$\mathcal{L}_{\text{int}}^{(2)} = A_\mu A^\mu \partial_\nu B^\nu \quad \rightarrow \quad \partial_\mu \phi \partial^\mu \phi \Box \psi + \ldots = \dot{\phi}^2 \ddot{\psi} - (\delta^{ij} \partial_i \phi \partial_j \phi) \ddot{\psi} + \ldots, \tag{A.45}$$

$$\mathcal{L}_{\text{int}}^{(3)} = A_\mu B^\mu \partial_\nu A^\nu \quad \rightarrow \quad \partial_\mu \phi \partial^\mu \psi \Box \phi + \ldots = \dot{\phi} \dot{\psi} \ddot{\phi} - (\delta^{ij} \partial_i \phi \partial_j \psi) \ddot{\phi} + \ldots, \tag{A.46}$$

where again we have stuck to the higher-derivative terms. In this case, the problematic term cannot be recast as a total derivative term, hence introducing third-order time derivatives in the field equations. Observe that there is a particular combination of $\mathcal{L}_{int}^{(2)}$ and $\mathcal{L}_{int}^{(3)}$ that cancels the problematic term (up to boundary terms).

A.7 Theory with Two Scalars and Second-Order Derivatives

In principle, since we have two fields with second-order time derivatives in the Lagrangian, we can expect the propagation of up to 4 degrees of freedom. As indicated by the hint, we will first calculate the Hessian associated to the given Lagrangian:

$$\mathcal{K} = \begin{pmatrix} \frac{\partial^2 \mathcal{L}}{\partial \ddot{\phi}^2} & \frac{\partial^2 \mathcal{L}}{\partial \ddot{\phi} \partial \ddot{\psi}} \\ \frac{\partial^2 \mathcal{L}}{\partial \ddot{\psi} \partial \ddot{\phi}} & \frac{\partial^2 \mathcal{L}}{\partial \ddot{\psi}^2} \end{pmatrix} = \begin{pmatrix} 1 & 1 \\ 1 & 1 \end{pmatrix}, \tag{A.47}$$

which is clearly degenerate (vanishing determinant). Hence, we know that it will propagate less than 4 degrees of freedom, which leads us to look for a possible simplification of the presented Lagrangian.

Using the expression for the square of a sum we can write the Lagrangian as

$$\mathcal{L} = \frac{1}{2}(\ddot{\phi} + \ddot{\psi})^2 - \frac{1}{2}m^2(\phi + \psi)^2 + m^2 \phi \psi. \tag{A.48}$$

Now, making the redefinition $\alpha = \phi + \psi$ and $\beta = \phi - \psi$ we arrive at

$$\mathcal{L} = \frac{1}{2}\ddot{\alpha}^2 - \frac{1}{2}m^2 \alpha^2 + \frac{1}{4}m^2(\alpha^2 - \beta^2), \tag{A.49}$$

which clearly propagates 2 degrees of freedom, one of which is a ghost. Both degrees of freedom are contained in just one dynamical scalar with second derivatives, α, since β appears as an auxiliary field whose equation enforces it to be trivial ($\beta = 0$).

References

1. A. Adams, N. Arkani-Hamed, S. Dubovsky, A. Nicolis, R. Rattazzi, Causality, analyticity and an IR obstruction to UV completion. JHEP **10**, 014 (2006)
2. K. Aoki, H. Motohashi, Ghost from constraints: a generalization of Ostrogradsky theorem. JCAP **08**, 026 (2020)
3. G. Arciniega, P. Bueno, P.A. Cano, J.D. Edelstein, R.A. Hennigar, L.G. Jaime, Geometric inflation. Phys. Lett. B **802**, 135242 (2020)
4. G. Arciniega, J.D. Edelstein, L.G. Jaime, Towards geometric inflation: the cubic case. Phys. Lett. B **802**, 135272 (2020)
5. N. Arkani-Hamed, H.-C. Cheng, M.A. Luty, S. Mukohyama, Ghost condensation and a consistent infrared modification of gravity. JHEP **05**, 074 (2004)
6. J. Beltrán Jiménez, A. Delhom, Ghosts in metric-affine higher order curvature gravity. Eur. Phys. J. C **79**(8), 656 (2019)
7. J. Beltrán Jiménez, A. Delhom, Instabilities in Metric-Affine Theories of Gravity. Eur. Phys. J. C **80**(6), 585 (2020)
8. J. Beltrán Jiménez, L. Heisenberg, T. Koivisto, Coincident general relativity. Phys. Rev. D **98**, 044048 (2018)
9. J. Beltrán Jiménez, L. Heisenberg, T.S Koivisto, S. Pekar, Cosmology in $f(Q)$ geometry. Phys. Rev. D **101**(10), 103507 (2020)
10. J. Beltrán Jiménez, L. Heisenberg, G.J. Olmo, D. Rubiera-Garcia, Born–Infeld inspired modifications of gravity. Phys. Rept. **727**, 1–129 (2018)
11. J. Beltrán Jiménez, A. Jiménez-Cano, On the strong coupling of Einsteinian cubic gravity and its generalisations. J. Cosmol. Astropart. Phys. **2021**(01), 069 (2021)
12. J. Beltrán Jiménez and F. J. Maldonado Torralba, Revisiting the stability of quadratic Poincaré gauge gravity. Eur. Phys. J. C **80**(7), 611 (2020)
13. C.M. Bender, P.D. Mannheim, No-ghost theorem for the fourth-order derivative Pais-Uhlenbeck oscillator model. Phys. Rev. Lett. **100**, 110402 (2008)
14. G.R. Bengochea, R. Ferraro, Dark torsion as the cosmic speed-up. Phys. Rev. D **79**, 124019 (2009)
15. M. Blagojević, B. Cvetković, General Poincaré gauge theory: Hamiltonian structure and particle spectrum. Phys. Rev. D **98**, 024014 (2018)
16. P. Bueno, P.A. Cano, Einsteinian cubic gravity. Phys. Rev. D **94**(10), 104005 (2016)
17. P. Bueno, P.A. Cano, R.A. Hennigar, On the stability of Einsteinian cubic gravity black holes in EFT. **6** (2023)
18. S.M. Carroll, M. Hoffman, M. Trodden, Can the dark energy equation-of-state parameter w be less than -1? Phys. Rev. D **68**, 023509 (2003)
19. T. Chen, M. Fasiello, E.A. Lim, A.J. Tolley, Higher derivative theories with constraints: exorcising Ostrogradski's Ghost. JCAP **02**, 042 (2013)

A. Delhom et al., *Instabilities in Field Theory*, SpringerBriefs in Physics, https://doi.org/10.1007/978-3-031-40433-7

20. A. Cisterna, N. Grandi, J. Oliva, On four-dimensional Einsteinian gravity, quasitopological gravity, cosmology and black holes. Phys. Lett. B 135435 (2020)
21. J.M. Cline, S. Jeon, G.D. Moore, The Phantom menaced: constraints on low-energy effective ghosts. Phys. Rev. D **70**, 043543 (2004)
22. G. Cognola, E. Elizalde, S. Nojiri, S.D. Odintsov, S. Zerbini, Dark energy in modified Gauss-Bonnet gravity: late-time acceleration and the hierarchy problem. Phys. Rev. D **73**, 084007 (2006)
23. T. Damour, S. Deser, J.G. McCarthy, Theoretical problems in nonsymmetric gravitational theory. Phys. Rev. D **45**, R3289–R3291 (1992)
24. T. Damour, S. Deser, J.G. McCarthy, Nonsymmetric gravity theories: inconsistencies and a cure. Phys. Rev. D **47**, 1541–1556 (1993)
25. A. De Felice, S. Tsujikawa, Solar system constraints on f(G) gravity models. Phys. Rev. D **80**, 063516 (2009)
26. A. De Felice, S. Tsujikawa, Excluding static and spherically symmetric black holes in Einsteinian cubic gravity **5** (2023)
27. A. Delhom, Theoretical and observational Aspecs in metric-affine gravity: a field theoretic perspective. Ph.D. thesis, Valencia U (2021)
28. S. Deser, J. Franklin, B. Tekin, Shortcuts to spherically symmetric solutions: a Cautionary note. Class. Quant. Grav. **21**, 5295–5296 (2004)
29. S. Deser, B. Tekin, Shortcuts to high symmetry solutions in gravitational theories. Class. Quant. Grav. **20**, 4877–4884 (2003)
30. V.E. Díez, M. Maier, J.A. Méndez-Zavaleta, M.T. Tehrani, Lagrangian constraint analysis of first-order classical field theories with an application to gravity. Phys. Rev. D **102**, 065015 (2020)
31. M.E. Fels, C.G. Torre, The Principle of symmetric criticality in general relativity. Class. Quant. Grav. **19**, 641–676 (2002)
32. A. Ganz, K. Noui, Reconsidering the Ostrogradsky theorem: higher-derivatives Lagrangians ghosts and degeneracy. Class. Quant. Grav. **38**(7), 075005 (2021)
33. J. Garriga, B. Shlaer, A. Vilenkin, Minkowski vacua can be metastable. JCAP **11**, 035 (2011)
34. A. Golovnev, T. Koivisto, Cosmological perturbations in modified teleparallel gravity models. JCAP **11**, 012 (2018)
35. L. Heisenberg, A systematic approach to generalisations of general relativity and their cosmological implications (2018), arXiv: 1807.01725
36. L. Heisenberg, Generalization of the proca action. J. Cosmol. Astropart. Phys. **2014**(05), 015–015 (2014). (May)
37. R. Hojman, C. Mukku, W.A. Sayed, Parity violation in metric torsion theories of gravitation. Phys. Rev. D **22**, 1915–1921 (1980)
38. S. Holst, Barbero's Hamiltonian derived from a generalized Hilbert-Palatini action. Phys. Rev. D **53**, 5966–5969 (1996)
39. J.B. Jiménez, A. Jiménez-Cano, On the physical viability of black hole solutions in Einsteinian cubic gravity and its generalisations **6** (2023)
40. A. Jiménez-Cano, Metric-affine Gauge theories of gravity. Foundations and new insights. Ph.D. thesis, Granada U., Theor. Phys. Astrophys. (2021)
41. A. Jiménez-Cano, F.J. Maldonado Torralba, Vector stability in quadratic metric-affine theories. JCAP **09**, 044 (2022)
42. A. Joyce, B. Jain, J. Khoury, M. Trodden, Beyond the cosmological standard model. Phys. Rept. **568**, 1–98 (2015)
43. R. Klein, D. Roest, Exorcising the Ostrogradsky ghost in coupled systems. JHEP **07**, 130 (2016)
44. T. Kobayashi, M. Yamaguchi, J. Yokoyama, Generalized G-inflation: inflation with the most general second-order field equations. Prog. Theor. Phys. **126**, 511–529 (2011)
45. B. Li, T.P. Sotiriou, J.D. Barrow, $f(T)$ gravity and local Lorentz invariance. Phys. Rev. D **83**, 064035 (2011)
46. D. Lovelock, Divergence-free tensorial concomitants. Aequat. Math. **4**(1), 127–138 (1970)
47. D. Lovelock, The Einstein tensor and its generalizations. J. Math. Phys. **12**, 498–501 (1971)

48. J.W. Moffat, New theory of gravitation. Phys. Rev. D **19**, 3554 (1979)
49. J.W. Moffat, Nonsymmetric gravitational theory. Phys. Lett. B **355**, 447–452 (1995)
50. R.C. Myers, B. Robinson, Black holes in quasi-topological gravity. JHEP **08**, 067 (2010)
51. S. Nojiri, S.D. Odintsov, M. Sasaki, Gauss Bonnet dark energy. Phys. Rev. D **71**, 123509 (2005)
52. J. Oliva, S. Ray, A new cubic theory of gravity in five dimensions: black hole, Birkhoff's theorem and C-function. Class. Quant. Grav. **27**, 225002 (2010)
53. M. Ostrogradski, Memoires sur les equations differentielles, relatives au probleme des isoperimetres. Mem. Acad. St. Petersbourg **6**(4), 385–517 (1850)
54. R.S. Palais, The principle of symmetric criticality. Commun. Math. Phys. **69**(1), 19–30 (1979)
55. E. Poisson, *A Relativist's Toolkit: The Mathematics of Black-hole Mechanics* (Cambridge University Press, 2004)
56. J.M. Pons, Ostrogradski's theorem for higher-order singular Lagrangians. Lett. Math. Phys. **17**(3), 181–189 (1989)
57. M.C. Pookkillath, A. De Felice, A.A. Starobinsky, Anisotropic instability in a higher order gravity theory **4** (2020)
58. V.A. Rubakov, The null energy condition and its violation. Phys. Usp. **57**, 128–142 (2014)
59. M. Sánchez, Remarks on the notion of global hyperbolicity. EAS Publ. Ser. **30**, 201–204 (2008)
60. M.D. Schwartz, *Quantum Field Theory and the Standard Model* (Cambridge University Press, 2014)
61. A. Smilga, Classical and quantum dynamics of higher-derivative systems. Int. J. Mod. Phys. A **32**(33), 1730025 (2017)
62. R.M. Wald, *General Relativity* (Chicago University Press, Chicago, 1984)
63. R.P. Woodard, The theorem of Ostrogradsky (2015)
64. R.P. Woodard, Avoiding dark energy with 1/r modifications of gravity. Lect. Notes Phys. **720**, 403–433 (2007)
65. H. Yo, J.M. Nester, Hamiltonian analysis of Poincare gauge theory scalar modes. Int. J. Mod. Phys. D **8**, 459–479 (1999)
66. H.-J. Yo, J.M. Nester, Hamiltonian analysis of Poincare gauge theory: higher spin modes. Int. J. Mod. Phys. D **11**, 747–780 (2002)

Printed in the United States
by Baker & Taylor Publisher Services